AF327804

Dimensional Analysis

Qing-Ming Tan

Dimensional Analysis

With Case Studies in Mechanics

 Springer

Qing-Ming Tan
The Institute of Mechanics
The Chinese Academy of Sciences
Beisihuanxi Road 15
100190 Beijing
People's Republic of China
e-mail: qmtan@hotmail.com

ISBN 978-3-642-19233-3   e-ISBN 978-3-642-19234-0

DOI 10.1007/978-3-642-19234-0

Springer Heidelberg Dordrecht London New York

Cover design: eStudio Calamar S.L.

Printed on acid-free paper

Springer is part of Springer Science+Business Media (www.springer.com)

Foreword

The book "Dimensional Analysis" by Professor Qing-Ming Tan is now published
by Springer. Although dimensional analysis is a traditional subject, its application
to natural phenomena and technical problems has never failed to yield surprising
results. Professor Tan clarifies the fundamental principles of dimensional analysis
with rigor and then goes on to provide many examples of application illustrating
the power of the technique. I believe the book would be welcome by a broad range
of readers and constitute a valuable contribution to the literature on science and
engineering education.

An outstanding feature of this book is that it provides ample and detailed
examples of how dimensional analysis is applied to solving problems in various
branches of mechanics. The formalism of the principle and technique of dimensional analysis may appear simple. But this could be deceiving, because dimensional analysis can be a powerful tool only when it is based on a sound and clear
understanding of the physical phenomenon at hand. Case studies and hands on
practices are needed in order to fully master the technique.

Another interesting feature of this book is that the author adopts the approach
that begins by separating the dependent variables from the independent variables
on the basis of a physical analysis of the problem. By doing so, the cause and effect
relationship between the variables is made clear. This procedure also helps reduce
effectively the number of variables to one dependent variable and a single set of
independent variables. Some books treat dimensional analysis without making
such distinctions. Mathematically there is nothing wrong, but the latter procedure
reduces dimension analysis to pure formalism and may generate a batch of
numbers devoid of a clear physical interpretation.

In recent years Professor Tan has been lecturing on dimensional analysis at the
University of Science and Technology of China, the Peking University, the Beijing
Institute of Technology, the Institute of Mechanics (CAS) and many other institutions. These lectures have been very well received by the students. This book in
its present form is based on Professor Tan's lecture notes repeatedly refined over
the years. Professor Tan is himself an ardent practitioner of dimensional analysis

and some of the contents of this book are actually taken from results of his research. This undoubtedly gives the book an additional depth.

The learning of dimensional analysis ought to be encouraged. As the author said, dimensional analysis is an effective tool in the exposition of natural phenomena and should be an integral component of the basic training of scientists and engineers. Hence, the publication of this book is both significant and timely.

I am pleased to see this book published and I like to recommend it to graduate students, scientific and technical workers, believing that all will benefit from studying it.

Beijing, February 2011   Zhemin Zheng

Preface

Brief History of the Subject

Roots of the study of dimensional analysis appeared in the nineteenth century. In 1822, J. B. J. Fourier wrote his classic book "Analytic Theory of Heat". In this book he presented a view of *dimensional homogeneity* in which natural phenomena were rigorously described using only dimensionally homogeneous equations. Fourier was the first to extend the concept of dimension from geometry to physics. In his view of homogeneity, changing the unit of measurement altered not only the magnitude of a certain quantity, but also the magnitude of other related quantities. Later researchers enlarged Fourier's work in modeling experiments that were designed to understand and predict characteristics of the real motion of objects and the magnitude of characteristic quantities involved.

Lord Rayleigh made outstanding contributions by applying dimensional analysis to problems in various fields. In 1871, Rayleigh made progress in scientifically describing the color of the sky. Later, he advanced the study of hydrodynamic stability and hydrodynamic drag, which influenced development of the aeronautical industry in the twentieth century. Rayleigh also used dimensional analysis to analyze the tune emitted from strings by wind, and clarified Strouhal's experimental results on Aeolian tones.

In 1883, O. Reynolds applied principles of dimensional analysis in modeling experiments related to fluid flow. Reynolds found that the difference between regular laminar flow and disordered turbulent flow depended on the magnitude of stream velocity as well as on the magnitude of the dimensionless number $\rho v d/\mu$ (ρ = fluid density, μ = fluid viscosity coefficient, d = inner tube diameter and v = mean stream velocity). This dimensionless number became widely known as Re, the Reynolds Number. If Re exceeds a certain critical value, a stream changes from laminar state to turbulent state. The pioneering work of Reynolds combined several factors influential in fluid flow in a tube to form a single dimensionless ratio and noticed that such a dimensionless number has a criterion value for predicting transition from laminar flow to turbulent flow.

In the twentieth century, dimensional analysis was applied widely to theoretical problems, engineering and technological problems, and became a focus of higher education. In 1904, *L. Prandtl* made careful observations and applied dimensional analysis to analyze viscous flow that passes a body. Prandtl eventually presented the theory of boundary layer, which gave a solid foundation for calculating frictional drag and ushered in the dawn of a new era in which the theory of fluid mechanics applied to aircraft design. In 1914, E. Buckingham presented evidence that physical law can be expressed by dimensionless quantities, each of which was noted as π. Such formulation provided the basis for modeling experiments and numerical simulations that are still widely used. In 1922, P. W. Bridgman identified Buckingham's formulation as the π theorem, even though *Fourier* had long before pointed out the essence of Buckingham's formulation. People in Fourier's time simply did not pay sufficient attention to his approach. In the 1940s, G. I. Taylor skillfully applied dimensional analysis to obtain the propagation law of explosion waves for the first atomic bomb and predict that bomb's *TNT yield*. In the latter half of the twentieth century, the significance of dimensional analysis became clear and there was no doubt that more and more applications in various fields would be found in future. Dimensional analysis even attracted interest from specialists in politics and economics, and became a subject in college study like the course Dimensional Analysis in Political Science.

Purpose of the Book

During my last year in college, my tutor Professor Hongsun Lin explained that dimensional analysis is very useful in physics, engineering, and especially in mechanics. At that time, my classmates and I were the first mechanics undergraduates in China and our teachers did not systematically teach dimensional analysis. We knew that the Reynolds Number was an important dimensionless number related to viscous flow, but we could not understand dimensional analysis. Since then, I have learnt dimensional analysis and used it. In a word, I could not have advanced in my career without understanding dimensional analysis because that principle provides the only way to solve complex problems when there is no available mathematical model. To deal with such problems, it is natural for me to apply the steps of dimensional analysis:

1. Analyze physical effects involved in the problem.
2. Select corresponding governing parameters.
3. Design experiments.
4. Analyze and synthesize experimental data.
5. Establish dimensionless relationships between cause and effect parameters.

In the 1970s, China's Minister of Machine Building Industry Hong Shen organized a group of specialists to edit a multi-volume reference text in Chinese that could serve as a handbook for mechanical engineering. The result was the publication of the "Mechanical Engineering Handbook" (in Chinese), in which C. M. Cheng (Zhemin Zheng) contributed to the study of dimensional analysis in a chapter on theoretical fundamentals "Similarity Theory and Modeling". Cheng's focus on dimensional analysis was fruitful, judging from eminent Hsue-Shen Tsien's observation on the importance of dimensional analysis in Explosion Mechanics and in other fields.

"Problems in the field of Explosion Mechanics are much more complicated than problems of classical Solid Mechanics or Fluid Mechanics, so it does not seem suitable to start with fundamental principles of mechanics in order to construct theoretical models for Explosion Mechanics. In recent years, however, the feasible way has been to conduct small-scale experimentation that should be accompanied by rigorous dimensional analysis, making it possible to extract empirical law from experimental results. This way has also been effective in the practice of scientific research in Mechanics in the past half century"—Hsue-Shen Tsien, preface to "Proceedings of Blasting in Rock and Soil" (1980).

"Looking back at the history of development of Mechanics, it can be said that the methods in Mechanics, such as dimensional analysis, similarity laws, etc., developed and shown to be effective through their service to engineering and technology, can also provide effective services to the development of the Natural Sciences"—Hsue-Shen Tsien, preface to "Mechanics and Production Construction" (1982).

Today, in the twenty-first century, commonly used models and equations still cannot deal with increasingly complex problems, making it more necessary than ever to apply dimensional analysis and to design appropriate experimentation to expose cruxes and clarify causality. Revealing fundamental physical characteristics of problems requires dimensional analysis as well as physical analysis.

Yet, regrettably, there is still a lack of study materials related to dimensional analysis for undergraduate and graduate students. For example, although a second edition of the "Mechanical Engineering Handbook" (in Chinese) in 1996 expanded treatment of dimensional analysis and included more materials, the handbook is primarily a reference for engineers and students and does not function well as a textbook. Few lecture courses in China or elsewhere explain the universality and effectiveness of dimensional analysis.

In recent years, I have tried to make up for the lack of information about dimensional analysis by lecturing at several institutions, including the Institute of Mechanics of the Chinese Academy of Sciences, the University of Science and Technology of China, the Peking University, the Beijing Institute of Technology and the Shanghai University. These lectures were greeted with significant interest in dimensional analysis and many audience members urged me to produce a textbook based on the contents of the lectures. In addition, my highly respected teachers Prof. Pei Li and Prof. Che-Min Cheng encouraged me to write a book.

This book covers basic concepts related to dimensional analysis, well-known applications in mechanics, applications useful in dealing with classical problems and research progress in explosion mechanics made by C. M. Cheng's research group over the past forty years. The book is intended for senior undergraduates and graduate students of physics and engineering, as well as for research scientists and engineers in relevant fields. While it is assumed that readers have a basic understanding of physics, complex mathematics is minimized. A case study approach is adopted in which problem analysis leads to application of dimensional analysis in a way that reveals physical mechanisms governing the problems being considered. Thus, sound analysis of causality in physical problems is linked to sound dimensional analysis.

Acknowledgment

I am greatly indebted to many people for their help in preparation of this book. Prof. Pei Li and Prof. Che-Min Cheng encouraged me to write and publish this English edition. Prof. Sturge Donald and Prof. Xiao-Lin Wang helped with the difficult task of polishing my writing in English. My thanks also go to Prof. Zhong Ling and Dr. Xu-Hui Zhang for drawing the figures.

I am grateful to Professor G. I. Barenblatt and the Cambridge University Press for permission to reproduce Fig. 3.4, which holds the copyright.

March 2011   Qing-Ming Tan

Contents

Chapter 1
Introduction

Dimension, dimensional quantities, dimensionless quantities, fundamental quantities, and *derived quantities* are defined. *Essentials of dimensional analysis* are shown using a problem related to a *simple pendulum.*

1.1 Preliminary Approaches

A series of physical quantities can describe natural phenomena and engineering problems such that the physical laws governing those phenomena and problems can be understood. Revealing those physical laws involves three steps:

Step 1. Classifying *physical quantities* of a given phenomenon or problem according to the *natures* of these physical quantities

Step 2. Finding correlations that connect the physical quantities

Step 3. Finding *causality* that connects the physical quantities

To determine *causality,* it is necessary to understand physical links and relations in a phenomenon or problem. Fundamental principles of physics may then be used to find *parameters of cause and effect* governing the phenomenon or problem. Parameters must be ranked according to importance, and only parameters in the same class can be compared in terms of *magnitude.* Deeper analysis means better results, so the analyst needs rich experience and resourcefulness in order to succeed. Trial and error is the usual way to achieve satisfactory results.

1.2 Dimension

In this book, *dimension* is used to express the essential nature of a quantity so that quantities can be grouped. For example, the quantities length, time, and mass vary in nature, so that they have various dimensions. It is important to distinguish *dimension* from *unit.* A *physical quantity* has a particular *nature* (i.e., particular

Q.-M. Tan, *Dimensional Analysis*, DOI: 10.1007/978-3-642-19234-0_1,

dimension) but a *unit* is a measure used to compare quantities. If two quantities have different natures (e.g., quantity of length and quantity of mass), dimensions of these quantities differ, cannot be compared profitably, and do not relate to unit. Conversely, if two quantities have similar dimensions (e.g., two lengths or two masses), *magnitudes* of these quantities can be compared profitably.

Comparing magnitudes of two quantities X_1 and X_2 with similar dimensions allows three possibilities:

$$1.\ \frac{X_1}{X_2} > 1,\ 2.\ \frac{X_1}{X_2} = 1,\ \text{and}\ 3.\ \frac{X_1}{X_2} < 1.$$

Denominator quantity X_2 is a viable standard unit for comparing X_1 and unit X_2 may be denoted as U:

$$X_1 = \frac{X_1}{X_2} \cdot X_2 = \frac{X_1}{X_2} \cdot U,$$

where ratio $\frac{X_1}{X_2}$ = exact magnitude of X_1 and $\frac{X_1}{X_2}$ = *dimensionless* pure number.

The above three possibilities may be written so that comparison unit = quantity X_1:

$$1.\ \frac{X_2}{X_1} < 1\quad 2.\ \frac{X_2}{X_1} = 1,\ \text{and}\ 3.\ \frac{X_2}{X_1} > 1.$$

Clearly, unit is a quantity used for comparison and is not the nature of a quantity or the dimension of a quantity.

1.3 Quantities: Dimensional, Dimensionless, Fundamental, and Derived

Physical quantities that relate to a phenomenon or problem can be *dimensional quantities* or *dimensionless quantities*. For dimensional quantities, *magnitude* depends on the *unit* selected (e.g., unit of length, unit of time, unit of mass, and unit of force). For dimensionless quantities, *magnitude* usually depends on ratio of two quantities with same dimensions (e.g., ratio of different lengths, ratio of different times, and ratio of different forces or ratio of different energies). Physical quantities related to a phenomenon or problem can be divided into two systems: *fundamental quantities* and *derived quantities*. A system of fundamental quantities is such that the dimensions of the quantities in this system are mutually independent. Another system of derived quantities is such that the dimension of every quantity in this system can be expressed by a combination of the dimensions of fundamental quantities in the problem.

1.4 Measurement of Physical Quantities

A physical law or principle is determined through experiment or theoretical work. Through analysis and synthesis, quantities related to physical problems are examined, classified, and related. Correctness is verified according to whether or not the deduction conforms to experiments and observations.

Describing the physical phenomena requires measuring each quantity involved. A certain quantity of the same nature is taken as an appropriate standard unit of measurement so that quantities can be measured and compared. For example, if length L is to be measured, length L_0 can be taken as a unit in order to compare L, producing the magnitude of L, which can be denoted as l, i.e.,

$$L/L_0 = l, \quad \text{or } L = lL_0. \tag{1.1}$$

Measurement lacks absolute precision because of error related to technique, apparatus or the carefulness of the operator. For example, a gauge made of iridoplatinum was used for a long time to define the length of one meter. Later, precision improved and the length of one meter was redefined as 1,650,763.73 times the wavelength of the orange line of the isotope krypton.

In general, *error* Δl can be estimated and it is usual to represent Δl as a positive number. If physical quantity A is measured using quantity U as a measurement unit, magnitude can be a, but exact value is a_0 and error is Δa (>0):

$$A = aU \tag{1.2}$$

and

$$a = a_0 \pm \Delta a, \, a/a_0 = 1 \pm \Delta a/a_0$$

Performing addition, subtraction, multiplication, and division on magnitudes a and b for quantities A and B, where $b = b_0 \pm \Delta b$ and $b/b_0 = 1 \pm \Delta b/b_0$, produces dimensional relationships and dimensionless relationships:

1. $$a + b = (a_0 + b_0) \pm (\Delta a + \Delta b),$$
$$(a + b)/(a_0 + b_0) = 1 \pm (\Delta a + \Delta b)/(a_0 + b_0),$$

2. $$a - b = (a_0 - b_0) \pm (\Delta a + \Delta b),$$
$$(a - b)/(a_0 - b_0) = 1 \pm (\Delta a + \Delta b)/(a_0 - b_0),$$

3. $$a \cdot b = (a_0 \cdot b_0) \pm (a_0\Delta b + b_0\Delta a),$$
$$(a \cdot b)/(a_0 \cdot b_0) = 1 \pm (\Delta a/a_0 + \Delta b/b_0),$$

4. $$a/b = (a_0/b_0) \pm (a_0\Delta b + b_0\Delta a)/b_0^2,$$
$$(a/b)/(a_0/b_0) = 1 \pm (\Delta a/a_0 + \Delta b/b_0).$$

1.5 The Simple Pendulum

To show how *dimensional analysis* can reveal the essentials and inherent causality of a problem, it is useful to consider an idealized pendulum consisting of a weightless string of fixed length l and a small sphere of mass m (Fig. 1.1). In this case, the sphere is attached to the lower end of the string and the upper end of the string is fixed to a ceiling. Due to *gravity*, the sphere with initial deviation angle α oscillates around the plumb line at the fixed point on the ceiling within a definite *period* T_p.

Several approximation assumptions apply in the case of a *simple pendulum*:

1. Mass of the string is assumed $<<$ mass of the small sphere m.
2. Deformation of the string is assumed length l.
3. Compared to gravity, aerodynamic drag is assumed to be negligible.

Clearly, oscillation period T_p depends on four *governing parameters*:

1. mass of small sphere m;
2. length of string l;
3. gravitational acceleration g;
4. initial deviation angle α, thus:

$$T_\mathrm{p} = f(m, l, g, \alpha). \tag{1.3}$$

The pendulum is a simple mechanical system. *Independent variables* of function f include three *fundamental quantities* having dimensions: m has dimension mass, l has dimension length, and g has dimension acceleration. A fourth independent variable α is dimensionless because an angle is defined by the ratio of two lengths. The dimension of *dependent variable* T_p is time, which can be expressed by combining the dimension of fundamental quantities l and g (length/acceleration)$^{1/2}$. Thus, T_p is *derived quantity*.

Fig. 1.1 A simple pendulum

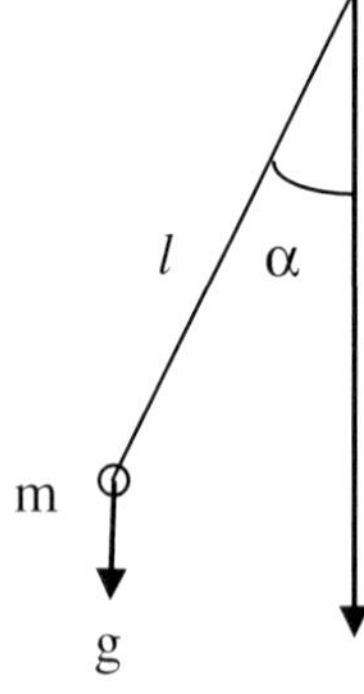

Fundamental quantities m, l, and g can serve as a unit system for measuring quantities related to the pendulum problem, so according to the relationship (1.3),

$$T_p \Big/ (l/g)^{1/2} = f(1,1,1,\alpha).$$

$T_p \big/ (l/g)^{1/2}$ varies only with α, so $T_p \big/ (l/g)^{1/2}$ is a function of α:

$$T_p \Big/ (l/g)^{1/2} = f_1(\alpha).$$

Discarding subscript "1" for function f_1 produces:

$$T_p \Big/ (l/g)^{1/2} = f(\alpha) \tag{1.4}$$

In expression (1.4), $f(\alpha)$ means that $T_p \big/ (l/g)^{1/2}$ is a function of α, not that $f(\alpha)$ is identical with $f(m,l,g,\alpha)$. Henceforth, f written without subscript indicates a functional relationship between dependent and independent variables, without specifying the particular function for such relationship.

Several conclusions appear:

1. T_p is proportional to $l^{1/2}$.
2. T_p is inversely proportional to $g^{1/2}$.
3. T_p does not depend on m.
4. $T_p \big/ (l/g)^{1/2}$ depends only on α. (The particular form of $f(\alpha)$ should be determined by experiment or theoretical analysis.)

Using direct experiments to seek the particular form of $T_p = f(m,l,g,\alpha)$ means that if each of the four independent variables requires ten experiments, then a total of 10^4 experiments are required. Using dimensional analysis, ten experiments can determine $f(\alpha)$. Furthermore, when initial deviation angle α is sufficiently small, i.e., $\alpha \ll 1$, the problem becomes much simpler. Judging from the physical point of view, $f(\alpha)$ must be an even function that can be expressed as a Taylor's series at $\alpha = 0$:

$$f(\alpha) = f(0) + f''(0) \cdot \alpha^2 + f^{(4)}(0) \cdot \alpha^4 / (4!) + \cdots \cong f(0)$$

There can also be approximate formulation of satisfactory precision:

$$T_p = (l/g)^{1/2} \cdot f(0) \tag{1.5}$$

A single experiment can determine the value of constant $f(0)$, but this constant is derived from theoretical analysis: $f(0) = 2\pi$.

1.6 Essential Principles

Constructing a model and deriving results for the example of the pendulum involves *essential principles* of dimensional analysis:

Principle 1. Only magnitudes of quantities of similar dimension can be compared.

In constructing a model, several important assumptions underlie this principle. For example, the mass of the string is assumed to be insignificant compared with the mass of the small sphere and string deformation is assumed to be insignificant compared with string length.

Principle 2. Physical phenomena and physical laws do not depend on the unit system selected.

Identifying a geometric figure provides a simple example of Principle 2. A triangle consists of sides l_1, l_2, and l_3 and regardless of the distance separating the triangle and an observer, shapes observed are similar because these shapes belong to the same class and any side of a triangle can be selected as a unit of length. For example, taking l_1 as a unit to measure, l_2 and l_3 provides magnitudes l_2/l_1 and l_3/l_1, respectively. Two such dimensionless quantities can be used to clarify these triangles, regardless of how far the observer is from what is observed. This method for identifying geometric figures can also be used to identify physical phenomena and to understand physical laws involved in these phenomena.

While the dimension of length is confined to geometric figures, dimensions related to physical quantities can not only include length but also time, mass and others, so a selection of fundamental dimensional quantities can serve as units for measuring variables related to physical phenomena. Properly selecting dimensionless quantities including dependent and independent variables, can be even more meaningful and essential than dimensional variables, allowing causality of a problem to be reformulated as a dimensionless expression. In multiple cases in the same class, not only in a few special cases, dimensionless causal relationship is more concise and more objective than dimensional relationship in reflecting the essentials of the physical phenomena.

Chapter 2
Fundamental Principles of Dimensional Analysis

A *power law formula* expresses the dimension of any quantity. There is proof of the Π theorem which is the theoretical foundation of *dimensional analysis* and the basis for deriving a principle of *similarity law* for modeling experiments and *numerical simulations*. Emphasis is placed on careful selection of fundamental quantities in solving problems.

2.1 Power Law Formula for Dimension of Physical Quantity

If unit U is selected to measure physical quantity X, magnitude of quantity x depends on unit selected:

$$X = xU \quad \text{and} \quad x = \frac{X}{U}. \tag{2.1}$$

Requirements apply to a mathematical operation that involve many physical quantities in a problem:

1. The operation must consider the dimension of the quantity examined and the relationship between that dimension and dimensions of fundamental quantities in the problem.
2. The operation must have a unified measurement unit system. To permit operation related to physical quantities, multiple unit systems must convert to a single unit system.

Generally, a problem in mechanics has *fundamental quantities* such as length, mass, and time and derived quantities such as velocity, density, and force. For example, if SI units and British units are selected simultaneously in measuring a certain quantity, magnitudes can be obtained in either unit system and there is a definite ratio between those two magnitudes.

Q.-M. Tan, *Dimensional Analysis*, DOI: 10.1007/978-3-642-19234-0_2,
© Springer-Verlag Berlin Heidelberg 2011

If SI units centimeter, gram, and second are used to measure length, mass, and time, then density of air under standard state $\rho = 0.129 \times 10^{-3} \text{g/cm}^3$, pressure $p = 1.013 \times 10^6 \text{g/(s}^2 \cdot \text{cm)}$ and speed of sound $c = 0.331 \times 10^5 \text{cm/s}$. If British units inch, pound, and second are used and compared with SI units, the unit of length is magnified by 2.54, mass is magnified by 453.6 and time is magnified by 1. Thus, magnitude of density ρ is $(2.54)^3/453.6$ (=0.0361), magnitude of pressure p is 2.54/453.6 (=5.60×10^{-3}), and magnitude of sound speed c in air of standard state is 1/2.54 (=0.394):

$$\rho = 1.29 \times 10^{-3} \times 0.0361 \text{lb/in.}^3 = 4.66 \times 10^{-5} \text{lb/in.}^3,$$

$$p = 1.013 \times 10^6 \times 5.60 \times 10^{-3} \text{lb/}\left(\text{s}^2 \cdot \text{in.}\right) = 0.567 \times 10^4 \text{lb/}\left(\text{s}^2 \cdot \text{in.}\right) = 4.7 \text{psi},$$

$$c = 0.331 \times 10^5 \times 0.394 \text{in./s} = 1.304 \times 10^4 \text{in./s} = 1.084 \times 10^3 \text{ft/s}.$$

In 1871, *J. C. Maxwell* [1] proposed that dimension of quantity X in a mechanics problem can be expressed by a *power law formula* in terms of dimensions of fundamental quantities length, mass, and time:

$$[X] = L^\alpha M^\beta T^\gamma, \tag{2.2}$$

where α, β, and γ are real numbers, and the dimension of any pure number (e.g., ϕ) that is irrelevant to the selection of the unit system can be formulated:

$$[\phi] = L^0 M^0 T^0 = 1. \tag{2.3}$$

If Maxwell is correct, in the unit conversion of any quantity X:

$$U \rightarrow U' \quad \text{and} \quad X = xU = x'U',$$

reduction ratio of magnitudes $\frac{x'}{x}$ follows corresponding power law formula:

$$\frac{x'}{x} = r_l^\alpha r_m^\beta r_t^\gamma, \tag{2.4}$$

where r_l, r_m, r_n = *reduction ratios* of the units length, mass, and time, respectively. Therefore, validating power law formulation (2.4) can be used to prove power law (2.2).

Maxwell's proposition can be proved in five steps:

Step 1. To measure physical quantity X, unit system U is selected, so $X = xU$.

Step 2. If unit system U' has reduction ratios for length, mass, and time r_l', r_m', and r_t', $X = xU = x'U'$. Ratio of magnitudes $\frac{x'}{x}$ depends on reduction ratios r_l', r_m', and r_t' and, without losing generality, $\frac{x'}{x}$ can be:

$$\frac{x'}{x} = f\left(r_l', \ r_m', \ r_t'\right). \tag{2.5}$$

To derive the particular form of function f:

Step 3. Another reduced unit system U'' has reduction ratios for length, mass, and time of r_l'', r_m'' and r_t'', so

$$X = x'U' = x''U'',\qquad(2.6)$$

Correspondingly, $\frac{x'}{x} = f\left(r_l',\ r_m',\ r_t'\right)$ and $\frac{x''}{x} = f\left(r_l'',\ r_m'',\ r_t''\right)$, so

$$\frac{x''}{x'} = \frac{f\left(r_l'',\ r_m'',\ r_t''\right)}{f\left(r_l',\ r_m',\ r_t'\right)}.\qquad(2.7)$$

Because reduction ratios for U' to U'' are $\frac{r_l''}{r_l'}$, $\frac{r_m''}{r_m'}$ and $\frac{r_t''}{r_t'}$, there is another relation:

$$\frac{x''}{x'} = f\left(\frac{r_l''}{r_l'},\ \frac{r_m''}{r_m'},\ \frac{r_t''}{r_t'}\right).\qquad(2.8)$$

Step 4. Comparing relations for $\frac{x''}{x'}$ in (2.7) and (2.8) allows the equation for the function of f:

$$\frac{f\left(r_l'',\ r_m'',\ r_t''\right)}{f\left(r_l',\ r_m',\ r_t'\right)} = f\left(\frac{r_l''}{r_l'},\ \frac{r_m''}{r_m'},\ \frac{r_t''}{r_t'}\right)\qquad(2.9)$$

If unit U' is fixed, unit U'' is changeable and U'' selection is arbitrary, super-script ($''$) can be omitted:

$$\frac{f\left(r_l,\ r_m,\ r_t\right)}{f\left(r_l',\ r_m',\ r_t'\right)} = f\left(\frac{r_l}{r_l'},\ \frac{r_m}{r_m'},\ \frac{r_t}{r_t'}\right).$$

Taking partial derivatives of both sides of the equation with respect to r_l and letting $(r_l,\ r_m,\ r_t)$ approach $(r_l',\ r_m',\ r_t')$ produces:

$$\frac{\frac{\partial f}{\partial r_l'}\left(r_l',\ r_m',\ r_t'\right)}{f\left(r_l',\ r_m',\ r_t'\right)} = \frac{1}{r_l'}\frac{\partial f}{\partial \frac{r_l}{r_l'}}(1,\ 1,\ 1),$$

Factor $\frac{\partial f}{\partial \frac{r_l}{r_l'}}(1,\ 1,\ 1)$ on the right hand side is a constant noted by α and U' is arbitrary, so superscript ($'$) can be omitted:

$$\frac{\frac{\partial f}{\partial r_l}\left(r_l,\ r_m,\ r_t\right)}{f\left(r_l,\ r_m,\ r_t\right)} = \frac{\alpha}{r_l}.\qquad(2.10a)$$

Similarly,

$$\frac{\frac{\partial f}{\partial r_m}\left(r_l,\ r_m,\ r_t\right)}{f\left(r_l,\ r_m,\ r_t\right)} = \frac{\alpha}{r_m},\qquad(2.10b)$$

and

$$\frac{\frac{\partial f}{\partial r_t}(r_l,\ r_m,\ r_t)}{f(r_l,\ r_m,\ r_t)} = \frac{\alpha}{r_t},\tag{2.10c}$$

where β and γ are constants.

Step 5. The integral is readily derived:

$$f(r_l,\ r_m,\ r_t) = r_l^{\alpha} r_m^{\beta} r_t^{\gamma}.\tag{2.11}$$

This derivation implies the corresponding expression for dimension of X:

$$[X] = L^{\alpha} M^{\beta} T^{\gamma}.\tag{2.12}$$

Clearly, when the unit system changes from U to U', reduction ratio of the magnitude of quantity $\frac{x'}{x}$ can be expressed by a power law monomial of the reduction ratios of units r_l, r_m, and r_t:

$$\frac{x'}{x} = r_l^{\alpha} r_m^{\beta} r_t^{\gamma}.\tag{2.13}$$

2.2 The Pi Theorem

In 1914, *E. Buckingham* presented the theoretical kernel of *dimensional analysis* in a theorem that denotes each dimensionless quantity as π, leading *P. W. Bridgman* to identify Buckingham's theorem as *the Pi theorem*. The idea central in proving the Pi theorem is similar to the idea used in discussing the simple pendulum in Chap. 1 of this book.

A definite function can express any physical law or principle. If physical phenomenon has n *independent variables* $a_1, a_2, \ldots, a_n$ and one *dependent variable* a, then a is a function of $a_1, a_2, \ldots, a_n$:

$$a = f(a_1,\ a_2, \ldots,\ a_k, a_{k+1},\ a_{k+2}, \ldots,\ a_n)\tag{2.14}$$

If the number of *fundamental quantities* is k, it is possible without losing generality to select k independent variables $a_1, a_2, \ldots, a_k$ with mutually independent dimensions as a group of fundamental quantities having dimensions $A_1, A_2, \ldots, A_k$, respectively. Remaining $n - k$ independent variables $a_{k+1}, a_{k+2}, \ldots, a_n$ are *derived quantities* with dimensions:

$$\begin{aligned}
1.\quad & [a_{k+1}] = A_1^{p_1} A_2^{p_2} \ldots A_k^{p_k}, \\
2.\quad & [a_{k+2}] = A_1^{q_1} A_2^{q_2} \ldots A_k^{q_k}, \\
& \qquad\qquad \vdots \\
n-k.\quad & [a_n] = A_1^{r_1} A_2^{r_2} \ldots A_k^{r_k};
\end{aligned}\tag{2.15}$$

where $p_1,\ldots, p_k$; $q_1,\ldots, q_k;\ldots$; $r_1,\ldots, r_k$ = relevant power values. Dependent variable a is *derived quantity* with dimension:

$$[a] = A_1^{m_1} A_2^{m_2} \ldots A_k^{m_k}, \tag{2.16}$$

where $m_1,\ldots, m_k$ are relevant power values.

Fundamental quantities $(a_1, a_2,\ldots, a_k)$ can be taken as a *unit system* to measure all variables in expression (2.14). Using expression (2.14), *magnitudes* of all the variables involved are dimensionless pure numbers:

$$\frac{a}{a_1^{m_1} a_2^{m_2} \ldots a_k^{m_k}} = f\left(1,\ 1,\ldots,\ 1;\ \frac{a_{k+1}}{a_1^{p_1} a_2^{p_2} \ldots a_k^{p_k}},\ \frac{a_{k+2}}{a_1^{q_1} a_2^{q_2} \ldots a_k^{q_k}},\ldots,\ \frac{a_n}{a_1^{r_1} a_2^{r_2} \ldots a_k^{r_k}}\right). \tag{2.17}$$

The left side of (2.17) $\frac{a}{a_1^{m_1} a_2^{m_2} \ldots a_k^{m_k}}$ is dimensionless dependent variable denoted as Π. On the right side, all magnitudes of first k independent variables $= 1$ and do not affect Π. Magnitudes of remaining $n - k$ independent variables denoted as Π_1, $\Pi_2,\ldots, \Pi_{n-k}$ determine magnitude of dependent variable Π, where

$$\Pi_1 = \frac{a_{k+1}}{a_1^{p_1} a_2^{p_2} \ldots a_k^{p_k}},\ \Pi_2 = \frac{a_{k+2}}{a_1^{q_1} a_2^{q_2} \ldots a_k^{q_k}}\ldots,\ \Pi_{n-k} = \frac{a_n}{a_1^{r_1} a_2^{r_2} \ldots a_k^{r_k}}.$$

In other words, dimensionless dependent variable Π is a definite function of $n-k$ *dimensionless independent variables*:

$$\Pi = f(\Pi_1,\ \Pi_2,\ldots,\ \Pi_{n-k}). \tag{2.18}$$

Variables Π_1, Π, $\ldots$, Π_{n-k} are mutually independent. Any of variables Π_i $(i = 1, 2, \ldots, n - k)$ can be replaced through combination with other variables Π_j $(j \neq i)$. For example, $\Pi_1' = \Pi_1^{\alpha_1} \Pi_2^{\alpha_2} \ldots \Pi_k^{\alpha_{n-k}}$ can replace Π_1, where $\alpha_1, \alpha_2,\ldots, \alpha_{n-k}$ are real numbers and $\alpha_1 \neq 0$.

Implicit function can be used to formulate the Pi theorem. If N variables $a_1, a_2,\ldots, a_N$ relate to a physical problem and include one dependent variable and N 1 independent variables, an implicit function can express the physical law of the problem:

$$f(a_1,\ a_2,\ldots,\ a_N) = 0. \tag{2.19a}$$

Without losing generality, k quantities $a_1, a_2,\ldots, a_k$ can be selected as fundamental quantities and the rest $N - k$ quantities can be selected as derived quantities. k fundamental quantities can be a unit system and (2.19a) can reduce to dimensionless relation:

$$f(1,\ 1,\ldots,\ 1;\ \Pi_1,\ \Pi_2,\ldots,\ \Pi_{N-k}) = 0, \tag{2.19b}$$

where Π_1, $\Pi_2,\ldots, \Pi_{N-k}$ = magnitudes of $a_{k+1}, a_{k+2},\ldots, a_N$ and magnitudes of the first k variables $a_1, a_2,\ldots, a_k = 1$ and do not affect function f. Because only $N - k$ dimensionless variables Π_1, $\Pi_2,\ldots, \Pi_{N-k}$ are significant, (2.19b) can be:

$$f(\Pi_1, \Pi_2, \ldots, \Pi_{N-k}) = 0. \tag{2.20}$$

To summarize the Pi theorem: if a physics problem has N variables that include one dependent variable and $N-1$ independent variables, k fundamental quantities and $N-k$ derived quantities, then $N-k$ dimensionless variables can form definite relationship that reflect the substance of the problem. The Pi theorem also shows that the physical law for a problem can be expressed as a *causal relationship* in dimensionless form.

2.3 Selection of Fundamental Quantities

Using dimensional analysis in a physics problem requires careful *selection of fundamental quantities* so that dimensions of these quantities are truly independent. A famous controversy about *heat conduction* in fluids between *Lord Rayleigh* and O. Riabouchisky shows the reason for such careful selection of fundamental quantities. The controversy suggests that reviewing a physics problem and obtaining a reasonable result requires grasping essentials of the problem from the outset and selecting proper fundamental quantities.

A paper published by Lord Rayleigh in 1915 [2] discusses *heat transfer* from inviscid fluid to solid body per unit of time (Fig. 2.1). The problem in Rayleigh's paper involves inviscid fluid with modest velocity v and temperature T_f that *flows past a solid body* with length l and temperature T_s, making temperature difference between fluid and solid body $T_f - T_s$, denoted as θ. In the problem, stream velocity causes modest pressure variation, so fluid is regarded as being incompressible. Fluid is also assumed to be inviscid, so neither compression nor viscosity effect in flow produces work that can be converted into thermal energy. In such a situation, *heat conduction* is solely responsible for creating a heterogeneous temperature field in which heat capacity c and conductivity coefficient λ of fluid must be considered. Fluid flow passes around and heats or cools the solid body and continues downstream. Clearly, rate of heat transfer H between fluid and solid body is a function of five governing parameters:

$$H = f(v, l, \theta, c, \lambda). \tag{2.21}$$

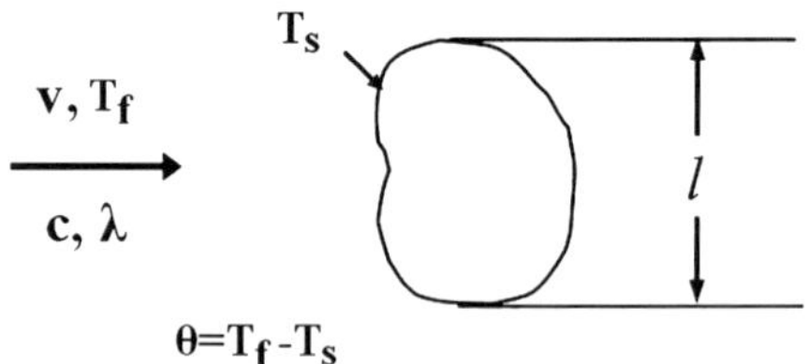

Fig. 2.1 Heat conduction in flow past a body

According to Rayleigh, independent dimensions in the problem are length L, time T, temperature K, and heat Q. Total number of physical quantities is $N = 6$ and the number of fundamental quantities is $k = 4$, so $N-k = 2$. According to the Pi theorem, two dimensionless quantities can form and generate dimensionless relationship:

$$H/(\lambda l\theta) = f\,(lvc/\lambda),$$

or

$$H = (\lambda l\theta) \times f\,(lvc/\lambda).$$

From this dimensionless relationship, Rayleigh deduces:

1. H is proportional to θ;
2. v and c do not influence H independently but combine to influence H with the product of v and c.

To argue against Rayleigh, *O. Riabouchinsky* [3] suggested that dimensions of temperature and heat are mutually dependent and identical to dimension of mechanical energy ML^2T^{-2} in the *L–M–T* system. For Riabouchinsky, independent dimensions in the problem are length L, mass M, and time T, i.e., $k = 3$, and $N - k = 3$ and three dimensionless quantities generate relationship:

$$H/(\lambda l\theta) = f\,(lvc/\lambda,\ cl^3). \tag{2.22}$$

According to Riabouchinsky, Rayleigh's first deduction is valid but his second deduction is invalid because v and c influence H independently.

Rayleigh [4] countered that he was following *Fourier's conduction law*, in which heat and temperature have mutually independent dimensions. According to Rayleigh, based on assumption that fluid is incompressible and inviscid, there is no transition mechanism between heat and mechanical energy in this heat conduction problem. For Rayleigh, temperature and heat are mutually independent dimensions and Riabouchinsky's point of view has no basis in physics.

Fourier's conduction equation $\frac{\partial T}{\partial t} = \frac{\lambda}{c}\left(\frac{\partial^2 T}{\partial x^2} + \frac{\partial^2 T}{\partial y^2} + \frac{\partial^2 T}{\partial z^2}\right)$ is the basis for Rayleigh's discussion and he focused on the fact that conductivity coefficient λ and heat capacity c do not appear independently in the equation for heat conduction but do appear in the form of ratio λ/c with dimension L^2/T. Fluid flowing over the solid body produces heat transfer and causes temperature difference ΔT between upstream temperature and downstream temperature. Thus, ΔT is a function of v, l, θ, and λ/c:

$$\Delta T = f\,(v, l, \theta, \lambda/c). \tag{2.23}$$

Rate of heat transfer H can be:

$$\int_{-\infty}^{+\infty} \int_{-\infty}^{+\infty} c \cdot \Delta T \cdot v\,\mathrm{dy}\,\mathrm{dz}, \tag{2.24}$$

where y and z are coordinates perpendicular to stream direction. Therefore:

$$H/(cv) = f(v, l, \theta, \lambda/c). \tag{2.25}$$

Independent dimensions in relation (2.25) are length L, time T and temperature K, so $k = 3$, $N = 5$, and $N-k = 2$, confirming Rayleigh's result:

$$H = (\lambda l \theta) \times f(lvc/\lambda). \tag{2.26}$$

2.4 Similarity Laws

Guided by the *Pi theorem*, an important principle can be drawn in formulating a physics problem: Presenting causal relationship in dimensionless form such as:

$$\Pi = f(\Pi_1, \Pi_2, \ldots, \Pi_{n-k})$$

is better than presenting causal relationship in dimensional form such as:

$$a = f(a_1, a_2, \ldots, a_k, a_{k+1}, \ldots, a_n),$$

where $a_1, a_2, \ldots, a_k$ = fundamental quantities, a and $a_{k+1}, \ldots, a_n$ = derived quantities, Π = dimensionless dependent variable corresponding to a, and $\Pi_1, \Pi_2, \ldots, \Pi_{n-k}$ = dimensionless independent variables corresponding to $a_{k+1}, \ldots, a_n$.

The best way to use theoretical analysis or *numerical simulation* in a physics problem is to start by properly selecting fundamental quantities that reflect the characteristics of the problem and then to take those fundamental quantities as the unit system for measurement. Solution is obtained in dimensionless form, e.g.:

$$\Pi = f(\Pi_1, \Pi_2, \ldots, \Pi_{n-k}). \tag{2.27}$$

Dependent variable Π can then be derived from given independent variables $\Pi_1, \Pi_2, \ldots, \Pi_{n-k}$.

The particular form of dimensionless causal relationship (2.27) can be obtained through modeling experiments or numerical simulation. If each independent variable in model (m), $(\Pi_1)_m, (\Pi_2)_m, \ldots, (\Pi_{n-k})_m$ = corresponding variable in prototype (p), $(\Pi_1)_p, (\Pi_2)_p, \ldots, (\Pi_{n-k})_p$:

$$(\Pi_1)_m = (\Pi_1)_p, \ (\Pi_2)_m = (\Pi_2)_p, \ldots, (\Pi_{n-k})_m = (\Pi_{n-k})_p, \tag{2.28}$$

Then dependent variable in model $(\Pi)_m$ = corresponding dependent variable in prototype $(\Pi)_p$:

$$(\Pi)_m = (\Pi)_p. \tag{2.29}$$

The *similarity law* (or *modeling law*) in (2.27) is a guiding principle for modeling experiments and *numerical simulation*. Dimensionless independent variables Π_1, Π_2, ..., Π_{n-k} are *similarity criterion numbers* or *similarity parameters*.

2.5 Applying the Pi Theorem

Several considerations apply to the use of the Pi theorem:

Consideration 1. In the causal relationship $a = f(a_1,\ a_2, \ldots,\ a_n)$ of a physics problem, the variables in function $f, a_1,\ a_2, \ldots,\ a_n$, must be the *independent variables*. There must be no dependent variable in the group of independent variables and there must be no irrelevant independent variable. It is necessary to estimate and compare how independent variables influence dependent variable in order to decide to include or exclude any independent variable. Every independent variable influences dependent variables, so deleting any independent variable destroys the integrity of the causal relationship, even if that variable is kept as a constant.

Consideration 2. The particular form of the function in dimensionless *causal relationship* $\Pi = f(\Pi_1,\ \Pi_2, \ldots,\ \Pi_{n-k})$ should be determined by experiment or theoretical analysis. Generally, that particular form cannot be obtained by relying only on the Pi theorem.

After processing experimental or numerical results, independent variables $(\Pi_1,\ \Pi_2, \ldots,\ \Pi_{n-k})$ can divide into domains and a power formula can be adopted that suits results in each domain, e.g.:

$$\Pi = c \cdot \Pi_1^{\alpha}\ \Pi_2^{\beta} \cdots \Pi_{n-k}^{\delta},$$

where $c = $ constant, and α, β,..., $\delta = $ real numbers. Logarithmic form:

$$\log\Pi = \log c + \alpha \cdot \log\Pi_1 + \beta \cdot \log\Pi_2 + \cdots + \delta \cdot \log\Pi_{n-k}. \tag{2.30}$$

Thus, in the double-logarithmic diagram, $\log\ c = $ intercept, and α, β,..., $\delta = $ slopes.

Consideration 3. Based on understanding of the nub of a physics problem, magnitude of the order of dimensionless independent variables Π_i can be analyzed. For example, if force F_1, F_2, and F_3 apply on a body, F_1 can be selected as a proper unit to produce dimensionless independent variables F_2/F_1 and F_3/F_1. If $F_3/F_1 \ll F_2/F_1$, then action of F_3 can be ignored and F_3/F_1 can be deleted from the group of dimensionless independent variables.

Consideration 4. Grasping the physical essence and applying a mathematical model to a physics problem is useful in finding the particular form of function f.

Consideration 5. For experiments or *numerical simulation*, it is better to start from causal relationship in dimensionless form $\Pi = f(\Pi_1,\ \Pi_2, \ldots,\ \Pi_{n-k})$ than to start from causal relationship in dimensional form $a = f(a_1,\ a_2, \ldots,\ a_n)$. Units do not need to be converted. The amount of work is considerably reduced because the number of independent variables is reduced and results are more universally significant.

References

1. Maxwell, J.C.A.: Treatise on Electricity and Magnetism. Clarendon Press, Cambridge (1871)
2. Rayleigh, L.: The principle of similitude. Nature **95**, 66–68 (1915)
3. Riabouchinsky, O.: The principle of similitude–letter to the editor. Nature **95**, 591 (1915)
4. Rayleigh, L.: The principle of similitude. Nature **95**, 644 (1915)

Chapter 3
Problems in Fluid Mechanics

Application of dimensional analysis to typical *problems in fluid mechanics* is introduced by using a case study approach, that consists of three steps: 1. analysis of basic physical principles involved in the problem, 2. introduction of governing parameters, and 3. making useful conclusions. Some commonly used similarity criterion numbers derived through mathematical formulation of fluid flow problems are identified. Other similarity criterion numbers for some special problems are given. Flow problems are classified according to flow conditions and corresponding similarity criterion numbers are identified.

3.1 Typical Flows

3.1.1 Overflow

In a general problem concerning *overflow* that passes a retaining wall, *gravity effect* governs overflow and discharge depends on water head and gravitational acceleration (Fig. 3.1).

It is assumed that storage capacity is large enough to have steady overflow and the section shape is an inverted triangle with bottom angle α or a rectangle with width b. Parameter density ρ indicates fluid *inertia* and parameters governing load applied to fluid are *gravity* acceleration g and water head h. These parameters can be used to determine mass discharge Q that passes the retaining wall:

$$Q = f(\rho, g, h, \alpha) \text{ or } Q = f(\rho, g, h, b). \tag{3.1}$$

In (3.1), dimensions are $[Q] = M/T$, $[\rho] = M/L^3$, $[g] = L/T^2$, $[h] = L$, $[b] = L$, and α is dimensionless.

This purely mechanical problem has three independent dimensions. *Selecting* ρ, g, and h as a *unit system* to measure variables creates dimensionless relationships:

$$\frac{Q}{\rho g^{1/2} h^{5/2}} = f(1,\ 1,\ 1,\ \alpha) \text{ or } \frac{Q}{\rho g^{1/2} h^{5/2}} = f\left(1,\ 1,\ 1,\ \frac{b}{h}\right).$$

Q.-M. Tan, *Dimensional Analysis*, DOI: 10.1007/978-3-642-19234-0_3,
© Springer-Verlag Berlin Heidelberg 2011

Fig. 3.1 Overflow

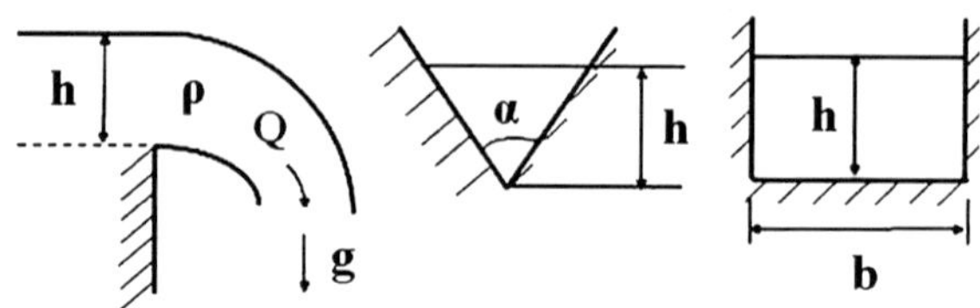

Fig. 3.2 Flow in tubes

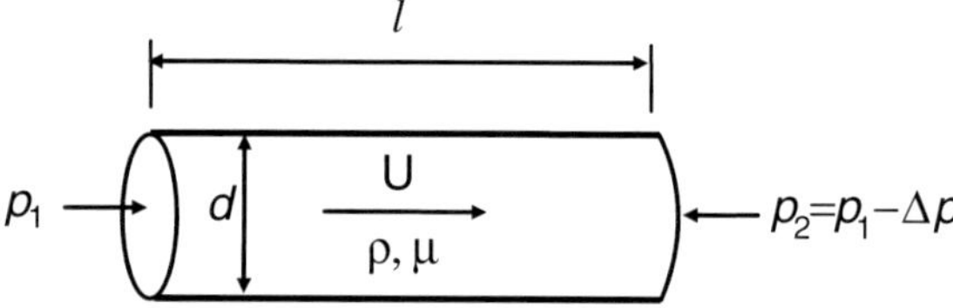

or:

$$\frac{Q}{\rho g^{1/2}h^{5/2}}=f(\alpha) \quad \text{or} \quad \frac{Q}{\rho g^{1/2}h^{5/2}}=f\left(\frac{b}{h}\right). \tag{3.2}$$

In the dimensionless relationships in (3.2), one independent variable replaces the four independent variables of the original dimensional relationships in (3.1). The value of similarity criterion number α or $\frac{b}{h}$ is sufficient to determine dimensionless relative discharge $\frac{Q}{\rho g^{1/2}h^{5/2}}$ and such relationships can apply to water, oil, and other fluids.

3.1.2 Friction Drag for Flows in Tubes

Flows in tubes are common in water supply, venting systems, heating systems, oil, or gas pipelines, etc. A key problem in engineering design and application is calculating correct tube size and pressure difference between inlet and outlet for given flow rate. Particularly in long distance transport, it is necessary to calculate distances between pumping stations and pressure differences that pumps must supply. *Viscosity* is important for such calculation, because flow drag relates to fluid viscosity and pressure difference overcomes *viscous drag* in flow.

It is assumed that a circular tube section with length l and inner diameter d transports incompressible *viscous fluid* with density ρ, *viscosity coefficient* μ and average section velocity U. The problem is to calculate pressure difference Δp between inlet and outlet (Fig. 3.2).

In terms of physics, pressure difference Δp must be proportional to tube length, so pressure difference for unit length $\Delta p/l$ should be a function of the tube's inner diameter, average velocity, fluid density, and fluid viscosity coefficient:

$$\frac{\Delta p}{l} = f(d, U, \rho, \mu). \tag{3.3}$$

Independent variables in (3.3) have three independent dimensions. *Selecting* $d, U,$ and ρ as a *unit system* and using the Pi theorem allows two dimensionless quantities to generate function relationship:

$$\frac{\Delta p}{\rho U^2} \cdot \frac{d}{l} = f\left(\frac{\mu}{\rho U d}\right),$$

or:

$$\frac{\Delta p}{\rho U^2} \cdot \frac{d}{l} = f\left(\frac{\rho U d}{\mu}\right), \tag{3.4}$$

On the right side of relationship (3.4), $\frac{\rho U d}{\mu}$ is *Reynolds number* Re that characterizes ratio of *inertia effect* to *viscosity effect*. The Reynolds number represents relative magnitude and is useful for comparing magnitude of tube inner diameter d with magnitude of characteristic length $\frac{\mu}{\rho U}$, magnitude of stream velocity U with magnitude of characteristic velocity $\frac{\mu}{\rho d}$ etc. Such characteristic length $\frac{\mu}{\rho U}$ and velocity $\frac{\mu}{\rho d}$ are inherent in this tube flow problem.

Dividing total drag by total inner tube wall area produces *friction drag* per tube wall unit area τ:

$$\tau = \frac{\Delta p \cdot \pi d^2/4}{\pi d \cdot l} = \Delta p \cdot \frac{d}{4l}. \tag{3.5}$$

Friction coefficient c_d may be defined as ratio of *friction drag* per unit area of tube wall τ to dynamic pressure $\frac{\rho U^2}{2}$:

$$c_d = \frac{\tau}{\rho U^2/2}, \tag{3.6}$$

which equals:

$$c_d = \frac{\Delta p}{\rho U^2/2} \cdot \frac{d}{4l}. \tag{3.7}$$

Considering (3.4), c_d is a function of Reynolds number:

$$c_d = f(\text{Re}), \tag{3.8a}$$

with detailed expression:

$$\frac{\Delta p}{\rho U^2/2} \cdot \frac{d}{4l} = f\left(\frac{\rho U d}{\mu}\right) \tag{3.8b}$$

Given sufficiently small Reynolds number, tube flow is regular and fluid particles move parallel to tube axis in what is called *laminar flow*. However, with sufficiently high Reynolds number, particles move irregularly. From the point of view of averaging space and time, fluid particles move from upstream to downstream. However, temporary velocity of those particles, including magnitude and direction, change from instant to instant in what is called *turbulent flow*.

In laminar flow, each fluid particle moves with constant velocity and fluid inertia does not influence flow, so the density that characterizes fluid inertia does not appear in friction coefficient expression $c_d = f(\mathrm{Re})$. The detailed form of this expression, $\frac{\Delta p}{\rho U^2/2} \cdot \frac{d}{4l} = f(\frac{\rho U d}{\mu})$, can reduce and function f is inversely proportional to $\rho U d/\mu$. Therefore:

$$\frac{\Delta p}{\rho U^2/2} \cdot \frac{d}{4l} = c \cdot \frac{\mu}{\rho U d}, \tag{3.9}$$

where c = proportion coefficient, and friction coefficient c_d is:

$$c_d = \frac{c}{\mathrm{Re}}. \tag{3.10}$$

Based on (3.9), pressure difference Δp can be:

$$\Delta p = c' \cdot \frac{\mu U l}{d^2}, \tag{3.11}$$

where c' = constant. Total drag for the tube is P:

$$P = \Delta p \cdot \frac{\pi d^2}{4} = c_1 \mu U l, \tag{3.12}$$

and friction force per unit area of wall τ is:

$$\tau = \Delta p \cdot \frac{d}{4l} = c_2 \frac{\mu U}{d}, \tag{3.13}$$

where c_1 and c_2 = constants.

Strictly speaking, the above discussion is appropriate only for a tube having sufficiently smooth inner wall. If the inner wall is not smooth, *roughness* must be a new parameter that can be defined as mean height of projections on the tube inner wall and denoted as k. *Relative roughness* can be defined by ratio of roughness k to tube radius $\frac{d}{2}$ and expressed as $\frac{k}{d/2}$.

Pressure difference required per tube unit length:

$$\frac{\Delta p}{l} = f(d, U, \rho, \mu, k), \tag{3.14}$$

where roughness k = additional parameter.

Three fundamental dimensions for independent variables remain and the Pi theorem permits generalized dimensionless relationship for friction coefficient c_d:

$$c_d = f\left(\frac{\rho U d}{\mu}, \frac{k}{d/2}\right). \tag{3.15}$$

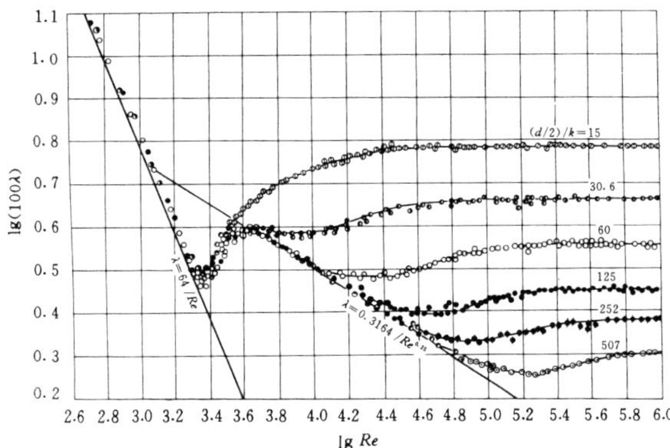

Fig. 3.3 Friction coefficient variation with Reynolds number and relative roughness

In 1932, *J. Nikuradze* applied $\lambda = 4c_d$ as a friction coefficient in experiments on various fluids and various circular tube sizes having various degrees of relative roughness. Experimental data was processed to form dimensionless relationship between friction coefficient and Reynolds number Re, in which relative roughness parameter $\frac{k}{d/2}$ is influential.

Figure 3.3 [1] shows that, experimental curves divide into *laminar regime*, *transition regime*, and *turbulent regime* as Reynolds number increases.

1. Laminar regime: For low Reynolds number, experimental points are on a single curve and friction coefficient decreases with increase of Re. The laminar regime conforms with preceding result $c_d = \frac{c}{\mathrm{Re}}$ and agrees with laminar relation $\lambda = \frac{64}{\mathrm{Re}}$ that is derived by theoretical hydrodynamic analysis.
2. Transition regime: With increase in Re, experimental points leave the laminar curve and spread into the turbulent regime, tending to curve according to varying value of relative roughness. At the start of transition from laminar regime to turbulent regime, Reynolds number is about 2×10^3.
3. Turbulent regime: The experimental curve agrees with varying value of relative roughness. With very low value of relative smoothness, a larger portion of the experimental curve overlaps the curve that corresponds to the smooth tube. For example, if relative roughness <0.2%, $\lambda \sim$ Re relationship coincides with the curve for a smooth tube $\lambda = \frac{0.3164}{\mathrm{Re}^{0.25}}$ in a large range for Re $< 10^5$.

    If there is significant roughness, experimental points are set on the $\lambda \sim$ Re relationship that corresponds to relative roughness. When the Reynolds number for given relative roughness surpasses a certain value, friction coefficient approaches a constant corresponding to relative roughness.

3.1.3 Velocity Profile of Viscous Flow in Cylindrical Tubes

Assuming that the tube inner wall is smooth, *roughness effect* can be ignored. To examine *velocity profile* along radial direction, the radial coordinate is regarded as an independent variable along with four parameters previously introduced (inner diameter d, mean velocity U, fluid density ρ, and viscosity μ). Here, distance perpendicular to the tube wall is the radial coordinate denoted as y. Velocity profile is:

$$\mathrm{u} = f(y;\ d,\ U,\ \rho, \mu), \tag{3.16}$$

where $u = $ velocity at y [2].

Selecting $d,\ U,\ \rho$ as a unit system allows dimensionless velocity profile:

$$\frac{u}{U} = f\left(\frac{y}{d}, \frac{\rho U d}{\mu}\right), \tag{3.17a}$$

or

$$\frac{u}{U} = f\left(\frac{\rho U y}{\mu}, \frac{\rho U d}{\mu}\right), \tag{3.17b}$$

where the second dimensionless number on the right hand side $\frac{\rho U d}{\mu} = $ Reynolds number for flow in the tube. Reynolds number magnitudes permit two cases related to flow:

Case 1. When Reynolds number is below a critical number, *laminar flow* occurs. Laws of kinematics govern fluid motion in that laminar flow, so fluid density does not influence flow. In other words, the function f for relative velocity u/U does not relate to density ρ, so both independent variables in the function appear as a ratio:

$$\frac{u}{U} = f\left(\frac{\rho U y}{\mu} \Big/ \frac{\rho U d}{\mu}\right)$$

and u/U can be:

$$\frac{u}{U} = f\left(\frac{y}{d}\right) \tag{3.18}$$

in a form validated by theoretical hydrodynamic analysis. The detailed form of the expression is:

$$\frac{u}{U} = 2 \cdot \left[1 - \left(\frac{r}{d/2}\right)^2\right], \tag{3.19}$$

where $r = $ distance from the tube axis and equals:

$$r = \frac{d}{2} - y. \tag{3.20}$$

Case 2. *Turbulent flow* occurs when Reynolds number is above a critical number and fluid particles move randomly. Dynamic mechanism of the flow is complicated, but flow relates to fluid density.

Regarding friction drag in a tube, *friction coefficient* c_d is a function of Reynolds number (see (3.8a) and (3.8b)):

$$c_d = \frac{\tau}{\rho U^2 / 2} = f(\mathrm{Re}), \qquad (3.21)$$

where τ = wall shear stress.

Assuming *wall shear stress* $\tau = \frac{\rho u_*^2}{2}$, where u_* = equivalent velocity characterizing wall friction, relative velocity $\frac{u_*}{U}$ is also a function of Re:

$$\frac{u_*}{U} = f(\mathrm{Re}). \qquad (3.22)$$

For given Re, relative velocity $\frac{u_*}{U}$ can be determined and u_* can replace U and represent characteristic velocity in the unit system. Therefore, velocity profile can be expressed by

$$u = f(y;\ d,\ U,\ \rho,\ \mu),$$

or:

$$u = f(y;\ d,\ u_*,\ \rho,\ \mu) \qquad (3.23)$$

Generally, ρ and μ act in the form of ratio $\frac{\mu}{\rho}$, which is called *kinetic viscosity* and denoted as v.

It can be assumed that:

$$\frac{\partial u}{\partial y} = f(y;\ d,\ u_*, v). \qquad (3.24)$$

In (3.24), two independent dimensions and five variables allow formation of three dimensionless quantities that are useful for generating dimensionless relationship:

$$\frac{y}{u_*} \frac{\partial u}{\partial y} = f\left(\frac{u_* y}{v};\ \frac{u_* d}{v}\right). \qquad (3.25)$$

Assuming that $\tilde{u} = \frac{u}{u_*}$ and $\eta = \frac{u_* y}{v}$:

$$\frac{\partial \tilde{u}}{\partial \ln \eta} = f\left(\eta;\ \frac{u_* d}{v}\right). \qquad (3.26)$$

Using various additional assumptions, various empirical laws for velocity profiles can be derived.

3.1.3.1 Empirical Law 1: *Logarithmic Law*

Assuming for fully developed turbulent flow, velocity gradient does not depend on molecular viscosity, i.e., $\partial \tilde{u}/\partial \ln \eta$ does not depend on dimensionless independent variables $\frac{u_* y}{v}$ and $\frac{u_* d}{v}$, so function f is constant:

$$\frac{\partial \tilde{u}}{\partial \ln \eta} = constant \tag{3.27}$$

Taking the integral produces:

$$\tilde{u} = c_1 \ln \eta + c_2, \tag{3.28}$$

where c_1 and $c_2 = $ integral constants. Expression (3.28) is exactly the logarithmic law given by *von Kármán* in 1930.

Combining empirical relationship and experimental results produces:

$$\frac{u}{u_*} = \frac{1}{\kappa} \ln \frac{y}{d/2}, \tag{3.29}$$

where *Kármán Constant* $\kappa \approx 0.425$.

3.1.3.2 Empirical Law 2: *Power-Type Law*

Based on a vast experimental data supplied by *J. Nikuradze*, *G. I. Barenblatt* assumed that viscosity slightly affects the intermediate region outside the tiny viscous layer near the tube wall and relatively far from the tube axis. For an observation point far from the wall, velocity distribution is approximately independent of the Reynolds number. Thus, dimensionless velocity gradient can be assumed:

$$f \sim A\eta^{\beta}, \tag{3.30}$$

where A and $\beta = $ asymptotic functions of parameter $\text{Re}\left(= \frac{u_* d}{v}\right)$ and both functions approach corresponding constants, provided $\text{Re} \rightarrow \infty$.

Thus:

$$\frac{\partial \tilde{u}}{\partial \ln \eta} = A\eta^{\beta}. \tag{3.31}$$

Taking the integral of expression (3.31) derives velocity distribution:

$$\tilde{u} = c\eta^{\alpha}, \tag{3.32}$$

where c and $\alpha = $ asymptotic functions of parameter Re.

Barenblatt used a fitting technique to obtain asymptotic relations for α and c:

$$\alpha = \frac{3}{2 \ln \text{Re}} \quad \text{and} \, c = \frac{1}{\sqrt{3}} \ln \text{Re} + \frac{5}{2}. \tag{3.33}$$

Thus, dimensionless velocity distribution $\frac{u}{u_*}$:

$$\frac{u}{u_*} = \left(\frac{1}{\sqrt{3}} \ln \text{Re} + \frac{5}{2}\right) \cdot \eta^{\frac{3}{2 \ln \text{Re}}}. \tag{3.34}$$

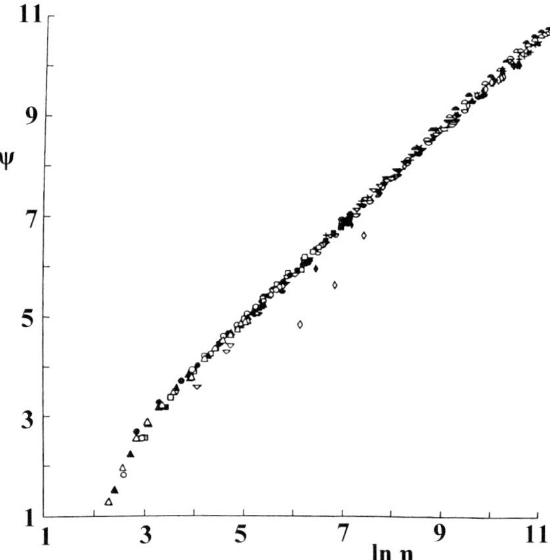

Fig. 3.4 Experimental points in reduced coordinates $(\psi, \ln \eta)$ (From [2])

Letting $\ln$ Re denote $3/(2\alpha)$, relationship (3.34) becomes:

$$\frac{1}{\alpha}\ln\left(\frac{2\alpha\tilde{u}}{\sqrt{3}+5\alpha}\right) = \ln \eta. \tag{3.35}$$

In the diagram of $\psi\left(= \frac{1}{\alpha}\ln\left(\frac{2\alpha\tilde{u}}{\sqrt{3}+5\alpha}\right)\right) \sim \ln\eta$, relationship (3.35) represents a bisector of the first quadrant. Inserting Nikuradze's experimental points for sixteen groups of different Reynolds number allows velocity distribution to be represented by that bisector in the intermediate region of $\ln\eta > 3$ (Fig. 3.4).

The asymptotic power-type law in (3.35) degenerates into logarithmic law when $\ln\eta$ tends to infinity. In geometric representation, logarithmic law relates closely to the envelope of a family of power-type curves and each curve corresponds to a fixed Reynolds number.

3.1.4 Flows Past a Body

Resistance is due to fluid *viscosity* in the motion of an aircraft in air or a submarine in water. Regardless of whether or not the solid body or the fluid serves as a reference coordinate, resistance is the same from the point of view of relative motion. Assuming the view of an observer moving together with the solid body with velocity v, fluid moves with the same magnitude of upstream velocity v but in the opposite direction and past the solid body (Fig. 3.5).

Assuming that characteristic lengths of solid body are $l, l',\ldots$, relative velocity between solid body and fluid $= v$, attack angle $= \alpha$ and velocity is so low that

Fig. 3.5 Flow past a body

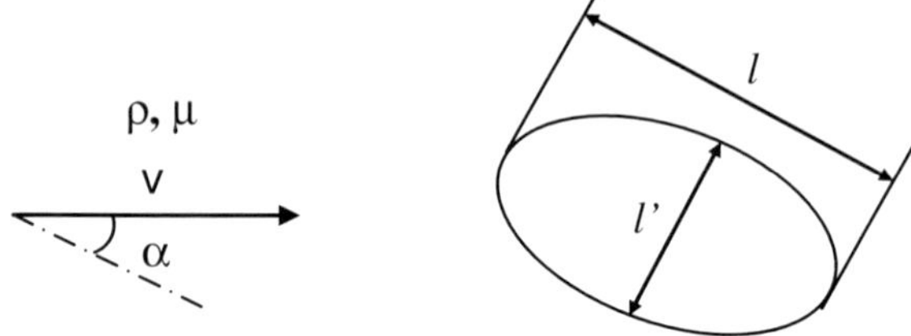

fluid *compressibility effect* can be ignored, additional governing parameters to be considered are fluid density ρ and viscosity μ (Fig. 3.5). Force W exerted on the solid body is a function of the above parameters:

$$W = f(l, \ l', \ \ldots; \ v, \ \alpha; \ \rho, \ \mu). \tag{3.36}$$

Selecting fundamental quantities l, v, and ρ as a unit system produces dimensionless relationship:

$$\frac{W}{\rho v^2 l^2} = f\left(\frac{l'}{l}, \ \ldots; \ \alpha; \ \mathrm{Re}\right) \quad \mathrm{Re} = \frac{\rho v l}{\mu}, \tag{3.37a}$$

or

$$W = \rho v^2 l^2 \cdot f\left(\frac{l'}{l}, \ \ldots; \ \alpha; \ \mathrm{Re}\right), \tag{3.37b}$$

where particular expression of function f is determined by experiment or theoretical analysis.

If *modeling experiments* determine dimensionless function f, dimensionless independent variables in the model (subscript m) equal those in the prototype (subscript p):

$$\left(\frac{l'}{l}, \ldots; \ \mathrm{Re}\right)_m = \left(\frac{l'}{l}, \ldots; \ \mathrm{Re}\right)_p, \tag{3.38}$$

then dimensionless dependent variables in model and prototype are equal:

$$\left(\frac{W}{\rho v^2 l^2}\right)_p = \left(\frac{W}{\rho v^2 l^2}\right)_m. \tag{3.39}$$

In other words, there are two conditions for modeling:

1. Geometric similarity:

$$\left(\frac{l'}{l}, \ldots\right)_m = \left(\frac{l'}{l}, \ldots\right)_p \quad \text{and } \alpha_m = \alpha_p;$$

Fig. 3.6 Resistance of a ship

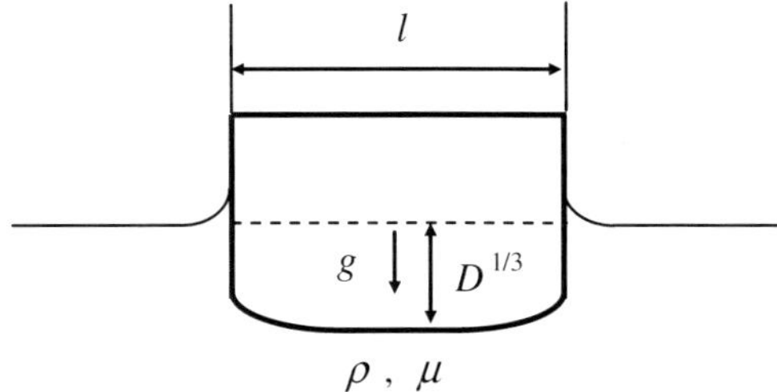

2. Dynamic similarity:

$$(\text{Re})_m = (\text{Re})_p.$$

If fluid used in the model is unlike that used in the prototype, generally $\left(\frac{\mu}{\rho}\right)_m \neq \left(\frac{\mu}{\rho}\right)_p$. To satisfy the similarity condition requires that $\frac{(vl)_m}{(vl)_p} = \frac{(\mu/\rho)_m}{(\mu/\rho)_p}$ in order to have:

$$\frac{W_m}{W_p} = \frac{(\rho v^2 l^2)_m}{(\rho v^2 l^2)_p} = \frac{(\mu^2/\rho)_m}{(\mu^2/\rho)_p}. \tag{3.40}$$

3.1.5 Resistance Related to Ships

Ships moving on the surface of water have two forms of *resistance*:

1. *Viscous resistance* is due to the relative motion of ship and water;
2. *Wave resistance* is due to moving ships producing energy dissipation of the surface wave.

Geometric factors influencing these two forms of resistance are characteristic length l that represents ship shape and displacement volume D that represents load weight. *Viscous resistance* also depends on ship velocity v and water properties that include density ρ and viscosity μ. In essence, wave resistance is a *gravity effect* that relates to gravitational acceleration g. (Fig. 3.6) Therefore, resistance of ship W is a function of the six parameters:

$$W = f(l,\ D;\ v;\ \rho,\ \mu,\ g). \tag{3.41}$$

The six independent variables in expression (3.41) contain three independent dimensions. Selecting l, v, and ρ as a unit system changes that dimensional relationship to a dimensionless relationship for relative resistance $\frac{W}{\rho v^2 l^2}$:

$$\frac{W}{\rho v^2 l^2} = f\left(\frac{D^{1/3}}{l}, \ \frac{\mu}{\rho v l}, \ \frac{g}{v^2/l}\right). \tag{3.42}$$

This dimensionless relationship in (3.42) contains three independent variables:

1. ratio of two lengths $\frac{D^{1/3}}{l}$ indicating relative load weight denoted as ψ;
2. reciprocal of *Reynolds number* (Re $= \frac{\rho v l}{\mu}$, ratio of *inertia effect* to *viscosity effect*);
3. reciprocal of *Froude number* (Fr $= \frac{v^2}{gl}$, ratio of inertia effect to *gravity effect*).
    Relationship (3.42) can be rewritten:

$$\frac{W}{\rho v^2 l^2} = f(\psi, \ \text{Re}, \ \text{Fr}). \tag{3.43}$$

Contact area between ship body and water or *wetting area* denoted as S depends on these six dimensional independent variables, so relative wetting area $\frac{S}{l^2}$:

$$\frac{S}{l^2} = f(\psi, \ \text{Re}, \ \text{Fr}), \tag{3.44}$$

where f has general meaning and no relation to the particular form of the function.

Ratio of resistance exerted on unit wetting area $\frac{W}{S}$ to dynamic pressure $\frac{\rho v^2}{2}$ defines resistant coefficient. This ratio is twice $\frac{W}{\rho v^2 l^2}\big/\frac{S}{l^2}$, so resistant coefficient:

$$\frac{W}{S}\bigg/\frac{\rho v^2}{2} = f(\psi, \ \text{Re}, \ \text{Fr}). \tag{3.45}$$

If small-scale modeling experiments are needed, three conditions are needed to match model and prototype:

$$(\psi)_m = (\psi)_p, \ (\text{Re})_m = (\text{Re})_p, \quad \text{and } (\text{Fr})_m = (\text{Fr})_p. \tag{3.46}$$

Relative resistance, relative wetting area and *resistance coefficient* obtained from modeling experiments should equal those dimensionless quantities in the prototype.

If fluid used in the model and prototype is the same and if experiments are carried out at the same height above sea level:

$$\left(\frac{\rho}{\mu}\right)_m = \left(\frac{\rho}{\mu}\right)_p, \quad \text{and } g_m = g_p. \tag{3.47}$$

To ensure necessary conditions related to Re and Fr:

$$(\text{Re})_m = (\text{Re})_p \quad \text{and } (\text{Fr})_m = (\text{Fr})_p, \tag{3.48}$$

requires:

$$(vl)_m = (vl)_p \quad \text{and} \quad \left(v^2/l\right)_m = \left(v^2/l\right)_p. \tag{3.49}$$

Thus, $l_m/l_p = 1$ must be satisfied and experimentation with scale models is worthless.

Due to requirements for simulating *viscosity effect* and *gravity effect*, ship-building engineers can seek only an approximate solution for ship resistance and assume that resistance of ship W is the linear summation of viscous resistance W_μ and wave resistance W_g:

$$W = W_\mu + W_g. \tag{3.50}$$

Furthermore, shipbuilding engineers assume that *viscous resistance* W_μ is proportional to total dynamic pressure $S \cdot \frac{\rho v^2}{2}$ and wave resistance W_g is proportional to total weight of displaced water $\rho g D$:

$$W = C_f(\mathrm{Re}) \cdot S \cdot \frac{\rho v^2}{2} + C_w(\psi,\ \mathrm{Fr}) \cdot \rho g D. \tag{3.51}$$

In relationship (3.51), the coefficient in viscous resistance C_f depends on Reynolds number and on shape and surface roughness of the ship. In practice, engineers use numerical simulations to obtain viscous resistance W_μ and use modeling experiments to determine total resistance W. Thus, the difference between W and W_μ is considered as wave resistance W_g.

3.1.6 Lubrication of Bearings

Bearings used in rotary machinery reduce friction and wear, so lubricating fluid is applied between axle and bearing. An axle and a bearing that rotate relative to one another produce equilibrium between fluid *viscous resistance* and pressure difference. Viscous resistance occurs on the surfaces of axle and bearing. Pressure difference is due to transfer of force from load borne by the axle to pressure distributed in fluid.

A problem related to bearing lubrication can be formulated directly or inversely. Direct formulation describes that an effect parameter such as bias α (ratio of bias distance to bearing diameter), friction drag F or flux of fluid Q is a function of several cause parameters for geometry and physics. Geometry parameters relate to the geometry of sliding surfaces, such as diameter of bearing R and diameter of axle r. The mean space between two sliding surfaces h is $R-r$. Physics parameters include *viscosity coefficient* μ of lubricating fluid, environmental pressure p_0 (in the case of sealing bearings), loading borne W and relative velocity v or angular velocity ω (Fig. 3.7).

In inverse formulation, bias α (or friction drag or fluid flux), load borne W, sliding velocity v, fluid viscosity μ and environmental pressure p_0 are given and geometry of axle and bearing is needed.

Fig. 3.7 Bearing lubrication

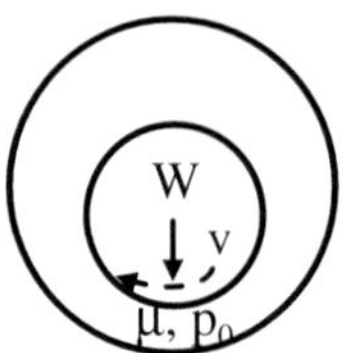

Inertia effect is not important because ratio of inertia force to viscous force can be $(\mathrm{Re})_h \cdot \frac{h}{R}$, where $(\mathrm{Re})_h = \frac{\rho v h}{\mu}$ and magnitude of order of $(\mathrm{Re})_h \cdot \frac{h}{R} < 10^{-3}$.

Whether or not the lubrication problem is examined directly or inversely, implicit function analysis may incorporate both direct and indirect formulation. All seven parameters can be regarded as being related variables having implicit function:

$$f(R,\ h,\ \mu,\ p_0,\ W,\ v,\ \alpha) = 0. \tag{3.52}$$

Taking R, p_0, and v as a unit system produces an implicit function equation having four dimensionless quantities:

$$f\left(\frac{h}{R},\ \frac{\mu v}{p_0 R},\ \frac{W}{p_0 R^2},\ \alpha\right) = 0, \tag{3.53a}$$

or

$$f\left(\frac{\mu \omega}{p_0},\ \frac{h}{R},\ \alpha,\ \frac{W}{p_0 R^2}\right) = 0, \tag{3.53b}$$

where the first dimensionless parameter $\frac{\mu \omega}{p_0}$ represents ratio of viscous stress to environmental pressure, the second and third are dimensionless geometrical parameters and the fourth represents ratio of loading pressure to environmental pressure.

In the case of gas-filled bearings, the *Harrison bearing number* denoted as Λ normally appears as a similarity criterion number that is defined by combining dimensionless parameters $\frac{\mu \omega}{p_0}$ and $\frac{h}{R}$:

$$\Lambda = \frac{\mu v}{p_0 R} \bigg/ \left(\frac{h}{R}\right)^2 = \frac{\mu v R}{p_0 h^2}. \tag{3.54}$$

Taking integrals for pressure distribution and viscous stress distribution allows calculation of load and friction drag operating on the bearing. For load W:

$$\frac{W}{p_0 R^2} = f_1\left(\Lambda,\ \frac{h}{R},\ \alpha\right). \tag{3.55}$$

For friction drag F:

$$\frac{F}{\mu v R} = f_2\left(\Lambda,\ \frac{h}{R},\ \alpha\right). \tag{3.56}$$

Thus, friction drag per unit load:

$$\frac{F}{W} = f_3\left(\Lambda,\ \frac{h}{R},\ \alpha\right). \tag{3.57}$$

Functions f_1, f_2, and f_3 in (3.55–3.57) are determined by experiments.

In modeling experiments, R_p/R_m is taken as the reduction ratio of length, denoted n, and the following conditions must be satisfied:

$$\Lambda_m = \Lambda_p,\ \left(\frac{h}{R}\right)_m = \left(\frac{h}{R}\right)_p,\quad \text{and } \alpha_m = \alpha_p. \tag{3.58}$$

Thus:

$$\left(\frac{W}{p_0R^2}\right)_p = \left(\frac{W}{p_0R^2}\right)_m\ \text{and}\ \left(\frac{F}{\mu vR}\right)_p = \left(\frac{F}{\mu vR}\right)_m. \tag{3.59}$$

If equal magnitude μv is used in model and prototype:

$$(\mu v)_m = (\mu v)_p. \tag{3.60}$$

Considering the requirement for similarity criterion number $\frac{\mu v}{p_0 R}$:

$$\left(\frac{\mu v}{p_0R}\right)_m = \left(\frac{\mu v}{p_0R}\right)_p, \tag{3.61}$$

Thus:

$$(p_0R)_m = (p_0R)_p. \tag{3.62}$$

Modeling produces:

$$\left(\frac{W}{p_0R^2}\right)_p = \left(\frac{W}{p_0R^2}\right)_m, $$

Therefore:

$$\left(\frac{W}{R}\right)_m = \left(\frac{W}{R}\right)_p. \tag{3.63}$$

Based on requirements (3.60–3.63) and modeling results, reduction relations can be summarized:

Prototype	R	h	W	p_0
Model	R/n	h/n	W/n	np_0

Two additional points are relevant:

1. If relative sliding velocity v or mean space h is so large that magnitude of order of $(Re)_h \cdot \frac{h}{R}$ approaches 10^{-1}, inertia effect must be considered.

Fig. 3.8 Water wave
propagation parameters

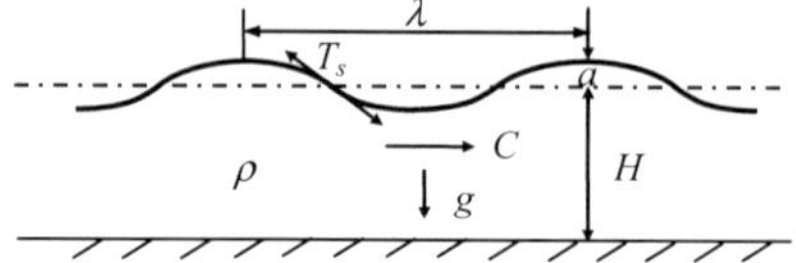

2. In the case of gas-filled bearings, if relative sliding velocity is high enough,
   effects of compressibility and thermal conductivity must be considered.

3.1.7 Water Waves

Water waves are phenomena of wave propagation on the surface of water.
Disturbed water particles near the surface displace from positions of equilibrium.
Because of inertia and restoring forces, particles vibrate around original equilibrium positions and cause continued disturbance that affects neighboring particles.
Vibrations and disturbances transmit from particle to particle and generate
wave motion. Generally, gravitational force is the main restoring force, but if
the wavelength of the water wave is sufficiently short, *surface tension* is also a
restoring force.

Wave amplitude of a water wave is generally small and pressure variation is
insufficient to change density significantly, so water can be regarded as being
incompressible. In such a case, three physical parameters governing propagation of
water waves are water density ρ (inertia), gravitational acceleration g (gravitation)
and surface tension T_s(surface force).

The simplest water waves are harmonic and geometric parameters are wavelength λ, wave amplitude a and water depth H.

Figure 3.8 shows all physical and geometric parameters that appear in a water
wave problem.

In *harmonic water waves* with sufficiently long *wavelength* and negligible
surface tension, *wave velocity c* is a function of five parameters:

$$c = f(\rho,\ g,\ \lambda,\ a,\ H), \tag{3.64}$$

Taking $\rho,\ g,$ and H as a unit system produces:

$$\frac{c}{\sqrt{gH}} = f\left(\frac{\lambda}{H},\ \frac{a}{H}\right). \tag{3.65}$$

Generally speaking, wave velocity varies with wavelength, so a water wave is
dispersive. A water wave composed of several elements of different wavelengths
changes waveform during propagation. In addition, wave velocity does not relate
to water density.

There are two extreme cases:

1. *Short waves in deep water*

In this case, $\lambda/H \ll 1$. Because depth influenced by water wave disturbance is less than wave length λ, water depth can be ignored. In addition, wave velocity c does not relate to water density, so:

$$c = f(g,\ \lambda,\ a). \tag{3.66}$$

Selecting g and λ as a unit system:

$$\frac{c}{\sqrt{g\lambda}} = f\left(\frac{a}{\lambda}\right). \tag{3.67}$$

For waves of small amplitude $a/\lambda \ll 1$, it is possible to approximate:

$$\frac{c}{\sqrt{g\lambda}} = k, \tag{3.68}$$

where $k = $ constant. Based on the theory of water waves, $k = \frac{1}{\sqrt{2\pi}}$.

2. *Long waves in shallow water*

If wavelength $>>$ water depth, the bottom significantly restricts vertical motion of water. Compared to water depth, wavelength effect is negligible in this case and waves are non-dispersive. In addition, wave density does not influence wave velocity, so:

$$c = f(g,\ H,\ a). \tag{3.69}$$

Selecting g and H as a unit system produces:

$$\frac{c}{\sqrt{gH}} = f\left(\frac{a}{H}\right). \tag{3.70}$$

For waves of small amplitude $a/H \ll 1$, it is possible to approximate:

$$\frac{c}{\sqrt{gH}} = k, \tag{3.71}$$

where $k = $ constant. The theory of water waves produces $k = 1$.

Harmonic waves with small wavelength are not only governed by gravity, but also governed by surface tension that cannot be ignored. Since the dimension of surface tension is $[T_s] = F/L = M/T^2$, dimensionless wave velocity c for harmonic wave of small amplitude can be expressed:

$$\frac{c}{\sqrt{gH}} = f\left(\frac{\lambda}{H},\ \frac{T_s}{\rho g \lambda^2}\right), \tag{3.72}$$

where $\frac{T_s}{\rho g \lambda^2} = $ reciprocal of the *Bond number*, which is generally defined:

$$\mathrm{Bo} = \frac{\rho g l^2}{T_s},$$

where $l = $ characteristic length.

Fig. 3.9 De Laval Nozzle

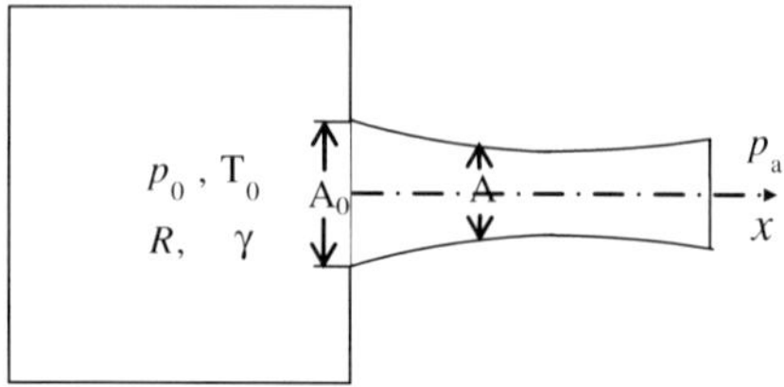

In the case of water wave problem, the characteristic length l = wavelength λ in harmonic waves, so from the viewpoint of physics, the Bond number = ratio of *gravity effect* to surface tension effect.

The particular form of function f in (3.73) is derived using the theory of water waves:

$$\frac{c}{\sqrt{gH}} = \sqrt{1 + \frac{(2\pi)^2 T_s}{\rho g \, \lambda^2}} \cdot \sqrt{\frac{\tanh(2\pi H / \lambda)}{2\pi H / \lambda}}. \tag{3.73}$$

The first factor on the right hand side of (3.73) indicates that, if $\frac{(2\pi)^2 T_s}{\rho g \lambda^2} \ll 1$ or wave length $\lambda \gg 2\pi \left(\frac{T_s}{\rho g}\right)^{1/2}$, surface tension can be ignored. For water, $T_s =$ 0.074N/m, $2\pi \left(\frac{T_s}{\rho g}\right)^{1/2} =$ approximately 1.7cm.

3.1.8 De Laval Nozzle

The *de Laval Nozzle* is a key rocket thruster component with a convergent-divergent channel. An entrance of the nozzle links to combustion chamber and an exit of the nozzle is open to the atmosphere. Combustion produces high pressure p_0 and high temperature T_0 in the combustion chamber, causing gaseous product to accelerate under pressure difference $p_0 - p_a$ (p_a = atmospheric pressure) in the nozzle and inject with high momentum into the atmosphere. Atmospheric environment applies inverse momentum to the nozzle, creating thrust to propel the rocket.

Using a one-dimensional approach to the above thrust process, nozzle axis = x axis and flow direction = positive direction of x. Based on the conservation of energy principle, the sum of internal energy and kinetic energy at every cross section is a constant equal to internal energy in the combustion chamber. During acceleration, kinetic energy of a particle increases, internal energy decreases, and total energy remains constant.

To consider a nozzle with cross sectional area $A(x)$ and cross section A_0 at the nozzle entrance, pressure of gaseous product in the chamber = p_0 and temperature of gaseous product in the chamber = T_0. Velocity in the chamber = 0 and environmental pressure = p_a (Fig. 3.9).

Assuming ideal gas with gas constant R and ratio of specific heat γ allows discussion of velocity distribution $v(x)$ and temperature distribution $T(x)$. Because of the distribution of cross section $= A(x)$, the velocity distribution can be expressed by $v(A)$ and the temperature distribution can be expressed by $T(A)$.

Generally, distributions of velocity and temperature relate to pressure p_a at nozzle exit. In a special case in which exit cross section area matches combustion chamber pressure, theoretical value of exit pressure $p_e =$ environmental pressure p_a, so p_a is not an independent governing parameter. In this special case, velocity and temperature distributions can be:

$$v = f_v(A_0,\ A;\ R,\ \gamma;\ p_0,\ T_0),\tag{3.74}$$

and

$$T = f_T(A_0,\ A;\ R,\ \gamma;\ p_0,\ T_0).\tag{3.75}$$

Mismatched cases $p_e \neq p_a$ are discussed at the end of this subsection.

In (3.75), *compressibility* is considered but viscosity is not considered, so the problem has four independent dimensions. Considering the seven variables in (3.74) or (3.75), three dimensionless independent variables can generate dimensionless function relationships:

$$\frac{v}{\sqrt{\gamma R T_0}} = f_v\left(\frac{A}{A_0},\ \gamma\right),\tag{3.76}$$

and

$$\frac{T}{T_0} = f_T\left(\frac{A}{A_0},\ \gamma\right).\tag{3.77}$$

Dividing (3.76) by (3.77):

$$\frac{v}{\sqrt{\gamma R T}} = f\left(\frac{A}{A_0},\ \gamma\right).\tag{3.78}$$

The left-hand-side of expression (3.78) $\frac{v}{\sqrt{\gamma R T}}$ is the *Mach number M* that reflects ratio of inertia to *compressibility*. Expression (3.78) can be rewritten as

$$M = f\left(\frac{A}{A_0},\ \gamma\right).\tag{3.79}$$

Expression (3.79) shows that M corresponds to $\frac{A}{A_0}$ for gas with given γ,

Gas dynamic theory provides the ordinary differential equation for function $M(A)$:

$$\frac{dM}{dA} = -\frac{1 + (\gamma - 1)M^2/2}{1 - M^2} \cdot \frac{dA/A}{A/A_0}.\tag{3.80}$$

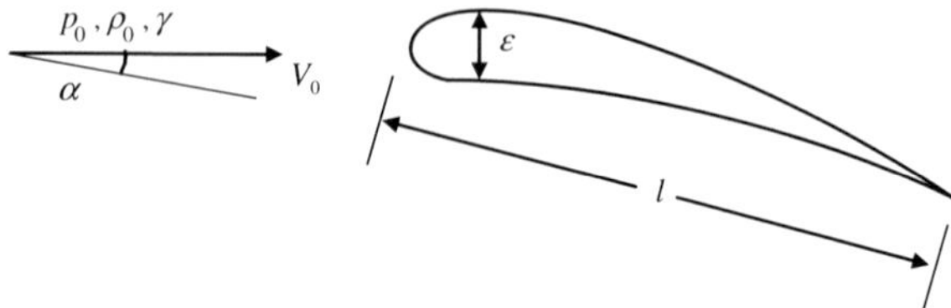

Fig. 3.10 High velocity flow past thin aerofoil

The right-hand-side of (3.80) shows factor $1 - M^2$ in the denominator. Clearly, *Mach number* variation for $M < 1$ in the subsonic region tends to be opposite of Mach number variation for $M > 1$ in the supersonic region.

To understand the process of continuous acceleration of the gas stream from stagnation state at nozzle entrance to supersonic state at nozzle exit requires separate analysis of tendency toward variation in two separate regions.

Because $M < 1$ in the earlier stage of acceleration, gas flow $\left(\frac{dM}{M} > 0\right)$ requires a convergent channel until velocity reaches sonic state $M = 1$, or $1 - M^2 = 0$.

To accelerate continuously and reach supersonic state ($M > 1$, or $1 - M^2 < 0$) requires a divergent channel $\left(\frac{dA}{A} > 0\right)$. Therefore, a nozzle with a convergent-divergent channel should power a supersonic jet.

If the shape of the divergent part does not match the thermodynamic state in the combustion chamber, theoretical value of exit pressure $p_e \neq$ environmental pressure p_a and there are two possibilities:

1. $\frac{p_e}{p_a} > 1$: Gas stream flows from the nozzle exit expands and accelerates continuously until pressure drops to p_a.
2. $\frac{p_e}{p_a} < 1$ The divergent part has no continuous flow. A shock wave occurs at a proper cross section in the divergent part, pressure jumps and velocity drops abruptly when gas stream passes through the shock wave. Pressure then increases continuously until reaching environmental pressure p_a at nozzle exit.

3.1.9 High Velocity Flow Past a Thin Aerofoil

In cases (3.1.1–3.1.8), simulation relates mainly to basic physical effects involved in problems and do not require complex mathematic analysis. The problem of high velocity *flow past a thin aerofoil* requires more complex mathematic analysis.

In the 1930s, interest in high velocity flows past a thin aerofoil prompted the scaling law for subsonic cases [3]. Upstream parameters are assumed: velocity $= v_0$, attack angle $= \alpha$, pressure $= p_0$, density $= \rho_0$, adiabatic index for air $= \gamma$ and air flows past a two-dimensional thin aerofoil. Aerofoil chord $= l$, characteristic thickness $= \varepsilon$. Flow is assumed to be irrotational, inviscid and dominated by *compressibility* effect (Fig. 3.10).

The two-dimensional velocity field (u, v) should be function of above governing parameters and coordinates x and y:

$$\begin{cases} u = f_u(x, y; v_0, \alpha, p_0, \rho_0, \gamma; l, \varepsilon) \\ v = f_v(x, y; v_0, \alpha, p_0, \rho_0, \gamma; l, \varepsilon) \end{cases}. \tag{3.81}$$

Taking v_0, ρ_0, and l as a unit system produces dimensionless relationships:

$$\begin{cases} \dfrac{u}{v_0} = g_u\left(\dfrac{x}{l}, \dfrac{y}{l}; \dfrac{v_0}{\sqrt{\gamma p_0/\rho_0}}, \alpha, \gamma; \dfrac{\varepsilon}{l}\right), \\ \dfrac{v}{v_0} = g_v\left(\dfrac{x}{l}, \dfrac{y}{l}; \dfrac{v_0}{\sqrt{\gamma p_0/\rho_0}}, \alpha, \gamma; \dfrac{\varepsilon}{l}\right). \end{cases} \tag{3.82}$$

Expression (3.82) has two dimensionless coordinates $\frac{x}{l}$, $\frac{y}{l}$ and four independent dimensionless parameters $\frac{v_0}{(\gamma p_0/\rho_0)^{1/2}}$, γ, α, and $\frac{\varepsilon}{l}$, where $(\gamma p_0/\rho_0)^{1/2} =$ sound velocity upstream a_0, $\frac{v_0}{(\gamma p_0/\rho_0)^{1/2}} = $ *Mach number* upstream, representing ratio of *inertia effect* to *compressibility effect*, $\alpha = $ attack angle, $\gamma = $ adiabatic index, and $\frac{\varepsilon}{l} = $ relative thickness of the aerofoil.

Because the thin aerofoil produces slight disturbance in the flow field, fundamental equations and boundary conditions controlling the flow field can be linearized.

Fundamental equations include equation of mass conservation, equation of momentum conservation and adiabatic relationship:

$$\begin{cases} \text{Mass conservation:} \qquad \dfrac{\partial \rho u}{\partial x} + \dfrac{\partial \rho v}{\partial y} = 0, \\[2mm] \text{Momentum conservation:} \; u\dfrac{\partial u}{\partial x} + v\dfrac{\partial u}{\partial y} = -\dfrac{1}{\rho}\dfrac{\partial p}{\partial x}, \\[2mm] \qquad\qquad\qquad\qquad\quad u\dfrac{\partial v}{\partial x} + v\dfrac{\partial v}{\partial y} = -\dfrac{1}{\rho}\dfrac{\partial p}{\partial y}, \\[2mm] \text{Adiabatic relationship:} \qquad \dfrac{p}{\rho^\gamma} = \dfrac{p_0}{\rho_0^\gamma}. \end{cases} \tag{3.83}$$

Applying adiabatic relationship to sound velocity a:

$$a = \sqrt{\frac{dp}{d\rho}} = \sqrt{\frac{\gamma p}{\rho}}. \tag{3.84}$$

Eliminating p and ρ in fundamental equations (3.83) allows an equation for (u, v) that includes sound velocity a:

$$(a^2 - u^2)\frac{\partial u}{\partial x} - uv\left(\frac{\partial v}{\partial x} + \frac{\partial u}{\partial y}\right) + (a^2 - v^2)\frac{\partial v}{\partial y} = 0. \tag{3.85}$$

Assuming that velocity field (u, v) is the sum of upstream velocity $(v_0, 0)$ and disturbance velocity (u', v') and assuming that sound velocity a is the sum of upstream a_0 and disturbance a':

$$u = v_0 + u' \quad (u' \ll v_0), \qquad v = 0 + v' \quad (v' \ll v_0),$$

$$a = a_0 + a' \quad (a' \ll a_0).$$

Substituting these assumptions into equation (3.85) and retaining terms related to first order small disturbances produces the linearized equation:

$$(a_0^2 - v_0^2)\frac{\partial u'}{\partial x} + a_0^2\frac{\partial v'}{\partial y} = 0. \tag{3.86}$$

Because flow is irrotational, velocity potential ϕ appears: $u = \frac{\partial \phi}{\partial x}$ and $v = \frac{\partial \phi}{\partial y}$. Similarly, because velocity potential ϕ is the sum of potential of upstream ϕ_0 and small disturbance ϕ', so:

$$\phi = \phi_0 + \phi' \quad (\phi' \ll \phi_0), \, u' = \frac{\partial \phi'}{\partial x}, \quad \text{and } v' = \frac{\partial \phi'}{\partial y}$$

Substituting into equation (3.86) produces equation for disturbance potential φ':

$$\frac{\partial^2 \phi'}{\partial x^2} + \frac{1}{1 - M_0^2}\frac{\partial^2 \phi'}{\partial y^2} = 0. \tag{3.87}$$

For simplicity, the following discussion relates to flow past a thin aerofoil with symmetrical shape and zero attack angle. However, asymmetrical cases can be discussed similarly. In a symmetrical flow field, boundary conditions divide into two parts:

1. On the aerofoil surface $y = \pm \varepsilon h\left(\frac{x}{l}\right)$, fluid velocity direction parallels aerofoil tangent;
2. At a position far from the aerofoil, disturbances $= 0$.

Mathematical formulation of the flow problem:

$$\left\{ \begin{array}{l} \text{Equation: } \dfrac{\partial^2 \phi'}{\partial x^2} + \dfrac{1}{1 - M_0^2}\dfrac{\partial^2 \phi'}{\partial y^2} = 0, \\[2ex] \text{Boundary conditions:} \\[1ex] \quad \text{At aerofoil surface: } y = \pm \varepsilon \cdot h\left(\dfrac{x}{l}\right): \quad \dfrac{dy}{dx} = \dfrac{v}{u}. \\[1ex] \qquad\qquad\qquad\quad \text{After linearization:} \\[1ex] \qquad\qquad y = 0\pm: \quad \dfrac{1}{v_0}\dfrac{\partial \phi'}{\partial y} = \pm\dfrac{\varepsilon}{l}\dfrac{dh}{d(x/l)}, \\[2ex] \quad \text{At infinity: } \dfrac{\sqrt{x^2 + y^2}}{l} \gg 1: \quad \dfrac{\partial \phi'}{\partial x} = \dfrac{\partial \phi'}{\partial y} = 0. \end{array} \right. \tag{3.88}$$

For subsonic flow past a thin aerofoil, v_0 and l can be a unit system. In the differential equation, ordinate y combines with coefficient $\sqrt{1 - M_0^2}$ to produce dimensionless dependent and independent variables:

$$\Phi = \frac{\phi'}{v_0 l}, \quad X = \frac{x}{l}, \quad Y = \frac{y\sqrt{1 - M_0^2}}{l}.$$
(3.89)

Dimensionless mathematical formulation of the problem:

$$
\begin{cases}
\text{Equation: } \dfrac{\partial^2 \Phi}{\partial X^2} + \dfrac{\partial^2 \Phi}{\partial Y^2} = 0, \\[2mm]
\text{Boundary conditions:} \\[2mm]
\text{At aerofoil surface: } 0 \le X \le 1 : \left(\dfrac{\partial \Phi}{\partial Y}\right)_{Y \to 0\pm} = \dfrac{\varepsilon}{l}\dfrac{1}{\sqrt{1 - M_0^2}}\dfrac{dh}{dX}, \\[2mm]
\text{At infinity: } (X^2 + Y^2) \gg 1 : \dfrac{\partial \Phi}{\partial X} = \dfrac{\partial \Phi}{\partial Y} = 0.
\end{cases}
$$
(3.90)

The only related independent parameter is the product of $\frac{\varepsilon}{l}$ and $\frac{1}{\sqrt{1-M_0^2}}$, i.e., $\frac{\varepsilon}{l}\frac{1}{\sqrt{1-M_0^2}}$.

Therefore:

$$\Phi = f\left(X, \ Y, \ \frac{\varepsilon}{l}\frac{1}{\sqrt{1 - M_0^2}}\right).$$
(3.91)

This similarity law for subsonic flows past a thin body is called *Gothert's rule* and two important points relate to its application:

1. In modeling experiments, airfoil shape does not have to be geometrically similar to prototype shape.
2. Incompressible flow can simulate compressible flow. However, because $M_0 = 0$ in incompressible flow, the similarity condition is:

$$\left(\frac{\varepsilon}{l}\right)_{imcompressible} = \left(\frac{\varepsilon}{l}\frac{1}{\sqrt{1 - M_0^2}}\right)_{compressible}.$$
(3.92)

3.1.10 Centrifugal Compressors

Particularly in *centrifugal compressors*, flow is three-dimensional and complicated, so modeling experiments are generally used to obtain design data. It is assumed that flow is steady, adiabatic, compressible and dominated by inertia, *viscosity* and *compressibility*. If model and prototype have similar shape, related parameter sets are:

Set 1. Characteristics of machinery: Length D and rotation per minute n;

Set 2. Property of medium: Adiabatic index γ, gas constant R, and *viscosity coefficient* μ;

Set 3. Characteristics of flow: Total pressure at inlet p_1^*, total pressure at outlet p_2^*, total temperature at inlet T_1^*, total temperature at outlet T_2^* and mass flow rate Q_m;

Set 4. Thermodynamic properties of machinery: Ratio of total pressure $\frac{p_2^*}{p_1^*}$, isentropic efficiency η^* and power N.

If independent variables $= D$, n; γ, R, μ; p_1^*, T_1^* and Q_m, dependent variables $= p_2^*$, T_2^*; η^* and N, selecting D, R, p_1^* and T_1^* as a unit system produces four dimensionless independent variables and four dimensionless dependent variables:

Independent variables:

$$\frac{nD}{\sqrt{RT_1^*}}, \quad \frac{Q_m\sqrt{RT_1^*}}{D^2 p_1^*}, \quad \frac{Dp_1^*}{\mu\sqrt{RT_1^*}}, \quad \text{and } \gamma;$$

Dependent variables:

$$\frac{p_2^*}{p_1^*}, \frac{T_2^* - T_1^*}{T_1^*}, \eta^* \quad \text{and} \frac{N}{D^2 p_1^*\sqrt{RT_1^*}}.$$

Thus:

$$\frac{p_2^*}{p_1^*} = f_1\left(\frac{nD}{\sqrt{RT_1^*}}, \frac{Q_m\sqrt{RT_1^*}}{D^2 p_1^*}, \frac{Dp_1^*}{\mu\sqrt{RT_1^*}}, \gamma\right), \tag{3.93a}$$

$$\frac{T_2^* - T_1^*}{T_1^*} = f_2\left(\frac{nD}{\sqrt{RT_1^*}}, \frac{Q_m\sqrt{RT_1^*}}{D^2 p_1^*}, \frac{Dp_1^*}{\mu\sqrt{RT_1^*}}, \gamma\right), \tag{3.93b}$$

$$\eta^* = f_3\left(\frac{nD}{\sqrt{RT_1^*}}, \frac{Q_m\sqrt{RT_1^*}}{D^2 p_1^*}, \frac{Dp_1^*}{\mu\sqrt{RT_1^*}}, \gamma\right), \tag{3.93c}$$

$$\frac{N}{D^2 p_1^*\sqrt{RT_1^*}} = f_4\left(\frac{nD}{\sqrt{RT_1^*}}, \frac{Q_m\sqrt{RT_1^*}}{D^2 p_1^*}, \frac{Dp_1^*}{\mu\sqrt{RT_1^*}}, \gamma\right). \tag{3.93d}$$

Functions f_1, f_2, f_3 and f_4 in (3.93a–3.93d) should be determined by modeling experiments that hold constant for dimensionless independent variables, but it is hard to ensure strict geometrical similarity. For example, it is hard to maintain similar geometry regarding the working precision of blades, impellers and relative space. Adiabatic index γ represents thermodynamic properties of the gas. Error associated with having air replace combustion product or vapor may be acceptable for subsonic flows, but adiabatic index must remain constant for multi-compressors or multi-turbines. The other three dimensionless independent variables correspond to the *Reynolds number* and *Mach number*, terms well known in aerodynamics:

$$\frac{nD}{\sqrt{RT_1^*}} = \frac{30}{\pi}\sqrt{\gamma\frac{T_1}{T_1^*}}M_{u1},$$

$$\frac{Q_m\sqrt{RT_1^*}}{D^2 p_1^*} = \frac{\pi}{4}\frac{p_1}{p_1^*}\sqrt{\gamma\frac{T_1}{T_1^*}}M_1,$$

$$\frac{Dp_1^*}{\mu\sqrt{RT_1^*}} = \frac{p_1^*}{p_1}\sqrt{\gamma\frac{T_1}{T_1^*}}\frac{\mathrm{Re}_1}{M_1},$$

where nD can be expressed by tangential velocity at inlet u_1, i.e., $nD = \frac{60}{2\pi}u_1$, $M_{u1} = $ inlet tangential Mach number and $M_1 = $ inlet Mach number, $\frac{T_1^*}{T_1} = 1 + \frac{(\gamma-1)}{2}M_1^2$ and $\frac{p_1^*}{p_1} = \left[1 + \frac{\gamma-1}{2}M_1^2\right]^{\frac{\gamma}{\gamma-1}}$. These three dimensionless independent variables correspond to the inlet Reynolds number and inlet Mach number and must be constants in modeling experiments. If $\mathrm{Re}_1 > $ critical value $(\mathrm{Re}_1)_{cr}$, flow is turbulent and machinery properties are generally considered to be unrelated to Reynolds number.

3.2 Similarity Criterion Numbers for Hydrodynamic Problems

Mathematical formulation for a flow problem consists of two parts:

1. Controlling equations;
2. Initial conditions and boundary conditions.

Controlling equations describe state variation of fluid particles in time and space. Kinematical state of particles is described by velocity (v_1, v_2, v_3). Thermodynamic state is described by two independent thermodynamic parameters, generally density ρ and temperature T. Other thermodynamic parameters can be derived from ρ and T by using constitutive relations of the fluids. Forces exerted on particles are pressure p, viscous stress τ_{ij} and volumetric force X_i (general gravitational force). In addition to mechanical action, fluid particles absorb or release heat throughout interfaces between particles and *heat flux rate* is denoted as q_i. Initial conditions indicate the state of the flow field when motion begins. Boundary conditions indicate the mechanical and thermodynamic environment that influences the fluid system [4].

Controlling equations consist of conservation relations and constitutive relations for the fluids.

Conservation relations:

1. *Conservation of Mass*:

$$\frac{\partial\rho}{\partial t} + \frac{\partial\rho v_1}{\partial x_1} + \frac{\partial\rho v_2}{\partial x_2} + \frac{\partial\rho v_3}{\partial x_3} = 0; \tag{3.94a}$$

2. *Conservation of momentum*:

$$
\begin{cases}
\dfrac{\partial v_1}{\partial t} + v_1\dfrac{\partial v_1}{\partial x_1} + v_2\dfrac{\partial v_1}{\partial x_2} + v_3\dfrac{\partial v_1}{\partial x_3} = -\dfrac{1}{\rho}\dfrac{\partial p}{\partial x_1} + X_1 + \dfrac{1}{\rho}\left(\dfrac{\partial \tau_{11}}{\partial x_1} + \dfrac{\partial \tau_{12}}{\partial x_2} + \dfrac{\partial \tau_{13}}{\partial x_3}\right), \\[2ex]
\dfrac{\partial v_2}{\partial t} + v_1\dfrac{\partial v_2}{\partial x_1} + v_2\dfrac{\partial v_2}{\partial x_2} + v_3\dfrac{\partial v_2}{\partial x_3} = -\dfrac{1}{\rho}\dfrac{\partial p}{\partial x_2} + X_2 + \dfrac{1}{\rho}\left(\dfrac{\partial \tau_{21}}{\partial x_1} + \dfrac{\partial \tau_{22}}{\partial x_2} + \dfrac{\partial \tau_{23}}{\partial x_3}\right), \\[2ex]
\dfrac{\partial v_3}{\partial t} + v_1\dfrac{\partial v_3}{\partial x_1} + v_2\dfrac{\partial v_3}{\partial x_2} + v_3\dfrac{\partial v_3}{\partial x_3} = -\dfrac{1}{\rho}\dfrac{\partial p}{\partial x_3} + X_3 + \dfrac{1}{\rho}\left(\dfrac{\partial \tau_{31}}{\partial x_1} + \dfrac{\partial \tau_{32}}{\partial x_2} + \dfrac{\partial \tau_{33}}{\partial x_3}\right).
\end{cases}
\tag{3.94b}
$$

3. *Conservation of energy*:

$$
\frac{d}{dt}\left(h + \frac{q^2}{2} + V\right) = \frac{1}{\rho}\frac{\partial p}{\partial t} + \frac{\partial V}{\partial t} - \frac{1}{\rho}\left(\frac{\partial q_1}{\partial x_1} + \frac{\partial q_2}{\partial x_2} + \frac{\partial q_3}{\partial x_3}\right) +
$$

$$
+ \frac{1}{\rho}\left(\frac{\partial(\tau_{11}v_1 + \tau_{12}v_2 + \tau_{13}v_3)}{\partial x_1} + \frac{\partial(\tau_{21}v_1 + \tau_{22}v_2 + \tau_{23}v_3)}{\partial x_2} + \frac{\partial(\tau_{31}v_1 + \tau_{32}v_2 + \tau_{33}v_3)}{\partial x_3}\right)
\tag{3.94c}
$$

In equation (3.94c), $\frac{d}{dt}$ = change rate revealed by following identified particle, $\frac{q^2}{2}$ = kinetic energy for fluid of unit mass, h = enthalpy for fluid of unit mass, i.e., sum of internal energy for fluid of unit mass u and $\frac{p}{\rho}$. V = potential of volumetric force X_i exerting on fluid unit mass $\left(X_i = -\frac{\partial V}{\partial x_i}\right)$.

Constitutive relations include equations for fluid state, heat conduction and Newtonian viscous stress:

1. *State equation*:

$$
p = p(\rho,\, T), \qquad u = u(\rho,\, T);
\tag{3.95a}
$$

2. *Heat conduction law*:

$$
q_i = -\lambda\frac{\partial T}{\partial x_i};
\tag{3.95b}
$$

3. *Viscous stress law*:

$$
\tau_{ij} = \mu\left(\frac{\partial v_i}{\partial x_j} + \frac{\partial v_i}{\partial x_j}\right) + \left(\mu' - \frac{2}{3}\mu\right)\cdot\delta_{ij}\cdot\left(\frac{\partial v_1}{\partial x_1} + \frac{\partial v_2}{\partial x_2} + \frac{\partial v_3}{\partial x_3}\right).
\tag{3.95c}
$$

where u = internal energy for unit mass fluid, q_i = rate of heat flow ($i = 1, 2, 3$), λ = thermal conductivity coefficient, τ_{ij} = viscous stress ($i = 1, 2, 3$, $j = 1, 2, 3$), μ = first *viscosity coefficient*, μ' = second *viscosity coefficient*, and δ_{ij} = Kronecker Symbol ($i = 1, 2, 3, j = 1, 2, 3$), i.e., $\delta_{ij} = 1$ when $i = j$ and $\delta_{ij} =$

0 when $i \neq j$. For incompressible fluids, viscous stress τ_{ij} does not relate to second viscosity coefficient μ': $\tau_{ij} = \mu(\frac{\partial v_i}{\partial x_j} + \frac{\partial v_j}{\partial x_i})$.

In steady flow, mathematical formulation includes controlling equations and boundary conditions. Kinematical and thermodynamic conditions at the boundary must be assigned for outer flow and inner flow. In general, boundary conditions are assigned at infinity and at solid body walls:

$$1.\ x_i \to \infty:\quad v_i = (v_0,\ 0,\ 0),$$
$$T = T_0\ ; \tag{3.96a}$$

$$2.\ x_i \to x_{iw}:\quad v_i = 0,$$
$$T = T_w(x_i)\ ;\quad \text{or}\ q_n = q_{nw}(x_i), \tag{3.96b}$$

where subscript "w" = solid body walls, and x_{iw} denotes position of those walls. Thermal condition at the walls = given temperature $T_w(x_i)$ or = given heat flux rate $q_{nw}(x_i)$ along the direction normal to the walls.

For gravitational force that acts as volumetric force, there are twelve governing parameters in mathematical formulations (3.94a–3.96a):

1. Characteristic length of solid body l;
2. Upstream velocity v_0;
3. Thermodynamic state of upstream ρ_0;
4. Thermodynamic state of upstream T_0;
5. Characteristic wall temperature T_{w0};
6. Characteristic wall *heat flux rate* q_{w0};
7. Thermal conductivity coefficient λ_0;
8. First viscosity coefficient of fluid μ_0;
9. Second viscosity coefficient of fluid μ'_0;
10. Specific heat at constant pressure c_{p0};
11. Specific heat at constant volume c_{v0};
12. Gravitational acceleration g,

where subscript "0" = upstream state.

Taking length, mass, and time as fundamental dimensions allows nine dimensionless similarity criterion numbers:

$$\left\{ \begin{array}{l} \dfrac{v_0^2}{gl} \equiv \mathrm{Fr},\ \dfrac{\rho_0 v_0 l}{\mu_0} \equiv \mathrm{Re},\ \dfrac{\mu_0 c_{p0}}{\lambda_0} \equiv \mathrm{Pr},\ \dfrac{v_0}{\sqrt{\gamma R_0 T_0}} \equiv M, \\[2ex] \dfrac{\mu'_0}{\mu_0},\ \dfrac{c_{p0}}{c_{v0}} \equiv \gamma,\ \dfrac{\rho_0 v_0^2 l^3}{T_0},\ \text{and}\ \dfrac{T_{w0}}{T_0}\ \text{or}\ \dfrac{q_{w0}}{\rho_0 v_0^3}. \end{array} \right. \tag{3.97}$$

Criterion numbers:

1. *Froude number* $\frac{v_0^2}{gl}$ (ratio of inertia to gravity);
2. *Reynolds number* $\frac{\rho_0 v_0 l}{\mu_0}$ (ratio of inertia to viscosity);

3. *Prandtl number* $\frac{\mu_0 c_{p0}}{\lambda_0}$ (ratio of viscosity to conductivity);

4. *Mach number* $\frac{v_0}{\sqrt{\gamma R_0 T_0}}$ (ratio of inertia to *compressibility*);

5. Ratio of two *viscosity coefficients* $\frac{\mu_0'}{\mu_0}$;

6. Ratio of two specific heats $\frac{c_{p0}}{c_{v0}}$;

7. Ratio of kinetic energy to thermal energy for upstream $\frac{\rho_0 v_0^2 l^3}{T_0}$;

8. Dimensionless characteristic temperature at walls $\frac{T_{w0}}{T_0}$;

9. Dimensionless characteristic *heat flux rate* at walls $\frac{q_{w0}}{\rho_0 v_0^3}$.

In *convective heat transfer*, it can be assumed generally that *heat flux rate* q_{nw} at the walls is proportional to $T_w - T_0$ and proportional coefficient α is *heat transfer coefficient*, so $q_{nw} = \alpha(T_w - T_0)$. Obtaining solutions related to distributions of velocity and temperature allows *heat transfer coefficient* α and relevant dimensionless parameter $\frac{\alpha l}{\lambda_0}$. *Nusselt number* $\frac{\alpha l}{\lambda_0}$ shows ratio of convective heat transfer to heat conduction, denoted as $\mathrm{Nu} = \frac{\alpha l}{\lambda_0}$ and commonly used as a dependent variable.

3.3 Additional Similarity Criterion Numbers

In addition to similarity criterion numbers in (3.97), other criterion numbers relate to various hydrodynamic problems.

1. *Rossby number*

In inner flows of *rotary machinery*, characteristic velocity v_0 can be understood as mean axial velocity. If angular velocity for rotation $= \omega$, dimensionless parameter $\omega l/v_0$ that usually relates to specific rotation n should be in fundamental equations. In problems related to weather or ocean currents, this parameter is the Rossby number, denoted Ro.

2. Conversion frequency

If, for example, flow is unsteady, fluids move past a solid body vibrating with frequency f and dimensionless parameter, *conversion frequency* fl/v_0 must be added.

3. *Grashof number*

In *natural convection* problems, there is no upstream velocity v_0 because thermal expansion effect and *gravity effect* cause convection. Volumetric force that produces convection is:

$$X_i = g\left(\frac{\rho}{\rho_0} - 1\right) \cdot n_i = g\beta\Delta T \cdot n_i = g\beta\Delta T_w \frac{\Delta T}{\Delta T_w} \cdot n_i,$$

where β = thermal expansion coefficient at constant pressure, n_i = unit vector and

$$\Delta T = T - T_0, \quad \Delta T_w = T_w - T_0.$$

Because all dimensions of X_i and $g\beta\Delta T_w$ are L/T^2, $\sqrt{gl\beta\Delta T_w}$ can replace v_0 in the Reynolds number and *Froude number*, where l is characteristic length. Thus, *Grashof number* Gr represents ratio of thermal buoyancy effect to viscous effect:

$$\mathrm{Gr} = \frac{gl\beta(T_w - T_0)}{[\mu_0/(\rho_0 l)]^2} = \frac{\rho_0^2 g l^3 \beta(T_w - T_0)}{\mu_0^2}.$$

4. Jacob number and Weber number

In problems of multi-phase flow with phase change such as *boiling* (vaporization) and condensation, latent heat for phase change H_0 and surface tension at bubble surface T_s should be considered, so the *Jacob number* and the Weber number are introduced.

The *Jacob number* represents ratio of fluid heat to heat supplied for phase change and is defined:

$$\mathrm{Ja} = \frac{c_p \rho_l(T_w - T_b)}{H_0 \rho_v},$$

where c_p = specific heat at constant pressure, ρ_l = density of liquid, T_w = temperature of bubble walls, T_b = bubble temperature and ρ_v = vapor density.

The *Weber number* represents ratio of inertia effect to surface tension effect and is defined:

$$\mathrm{We} = \frac{\rho_l v_b^2 D_b}{T_s},$$

where v_b = relative velocity of a bubble to surrounding liquid and D_b = bubble diameter.

Boiling number Bo is commonly used as a dependent variable to replace *Nusselt number* Nu and is defined:

$$\mathrm{Bo} = \frac{q}{H_0 \rho_v v_b},$$

where q − *heat flux rate* and boiling number represents ratio of heat flux effect to heat effect for phase change.

5. *Cavitation number*

In problems related to hydraulic machinery, e.g., propulsion by screw propellers over or under water surface and hydrofoils, cavitation is an important factor in work efficiency.

According to the Bernoulli theorem, flow pressure decreases as velocity increases. Bubbles occur when pressure in liquid is lower than vaporization pressure. If vaporization pressure = p_c, dimensionless parameter *cavitation number* σ needs to be introduced. This cavitation number is defined:

$$\sigma = \frac{p_0 - p_c}{\rho_0 v_0^2/2}.$$

Table 3.1 Main similarity criterion numbers in typical cases

Conditions	Similarity criterion numbers	Remarks
1. Incompressible fluid ($M \approx 0$)	Fr	Gravitation effect
	Re	Viscous resistant effect
		Re, very small—Stokes flow
		Re, small—Laminar flow
		Re, large—Turbulent flow
	$\omega l/v_0$	Rotation effect in rotary machinery
	σ	Cavitation phenomena
	Euler number: $Eu = \Delta p/(\rho v_0^2/2)$	Commonly used as dependent variable
2. Compressible flow	Fr, Re, $\omega l/v_0$, σ, Eu	Idem
	M	*Compressibility effect*
		$M < 0.3$—Negligible
		$0.3 < M < 1$—Subsonic flow
		$M \approx 1$—Transonic flow
		$M > 1$—Supersonic flow
		$M \gg 1$—Hypersonic flow, strong interference of boundary layer with flow behind shock wave, Re, Pr, and Nu should be considered and *heat transfer* related to ionization, erosion, and chemical reaction should be considered
	Pr	Pr ≈ 1 for air
	Nu	Commonly used as dependent variable
3. Heat transfer in *natural convection*	$\omega l/v_0$, σ, Eu	σ and Eu are used as dependent variables in the case of cavitation
	Gr	Buoyancy and *viscosity effects*
	Pr	Pr ≈ 1 for air
	Nu	Commonly used as dependent variable
4. Heat transfer in forced convection	Re, $\omega l/v_0$, σ, Eu	σ and Eu are used as dependent variables in the case of cavitation
	Pr	Pr ≈ 1 for air
	Nu	Commonly used as dependent variable
	M	Considered for $M > 0.3$
5. Heat transfer for *boiling*	Gr, Re, $\omega l/v_0$, σ, Eu	σ and Eu are used as dependent variables in the case of cavitation
	Pr	Pr ≈ 1 for air
	Ja	Phase transition induced thermal effect
	We	Surface tension effect
	Nu or Bo	Commonly used as dependent variable

3.4 Classification of Fluid Flow

Section 3.2 introduces numerous similarity criterion numbers through the use of mathematical formulation of common flow problems. In practice, solving a specific problem requires simplification according to conditions specific to that problem. Table 3.1 shows several main similarity criterion numbers in some typical flows. Geometrical similarity parameters are not included and there is no mention of complex problems such as multi-phage flows, flows with chemical reaction, flows with heat radiation or magneto-hydrodynamic flows. Solving such problems requires having references from specialized fields.

References

1. Loitsyanskii, L.G.: *Mechanics of Liquids and Gases*, Pergamon Press, Oxford, (1966) (Лойцянский, Л. Г.: *Механика Жидкости и Газа*. Гос. Изд. Техн-Теор Литературы, Москва, Ленинград, (1950))
2. *Barenblatt, G. I.: Scaling, Self-similarity, and Intermediate Asymptotics*. Cambridge, London. *Dimensional Analysis and Intermediate Asymptotics* (Cambridge Texts in Applied Mathematics) (Paperback) (1991, 1993)
3. Shapiro, A.H.: The Dynamics and Thermodynamics of Compressible Fluid Flow, vol. I, pp. 315–327. Ronald Press Co, New York (1953)
4. Tsien, H.S.: The equations of gas dynamics. In: Emmons, H.W. (ed.) Fundamentals of Gas Dynamics. Oxford University Press, London (1958)

Chapter 4
Problems in Solid Mechanics

Deformation, motion, and fracture of solids are discussed, including analysis of some typical problems and an introduction of recent progress in solving several problems in solid mechanics. There are discussion of stress analysis, stability analysis, vibration, and waves of elastic bodies, discussion of stress analysis of elasto-plastic bodies, as well as discussion of fracture analysis of solid bodies. Dimensional analysis is applied throughout a case study approach that analyzes physical principles involved in problems, introduces governing parameters and presents useful conclusions. Contents are arranged by order of difficulty and problems advance from one-dimensional to three-dimensional, from elastic to elasto-plastic and from static to dynamic.

4.1 Stress Analysis for Elastic Bodies and Stability Analysis for Simple Structures

Dimensional analysis can be applied to the distribution of deformation and stress in simple structures like elastic beams or plates, to three-dimensional elastic bodies as well as to the stability criterion of an Eulerian column under compression.

4.1.1 Deflection of Beams

Geometric similarity is not necessary for modeling some problems in solid mechanics, such as a problem related to simply-supported elastic beam (Fig. 4.1). For distributed loading $q(x)$ applied on the beam of unit length, deflection of the simply-supported beam $w = w(x)$ satisfies the following equation and boundary conditions:

Q.-M. Tan, *Dimensional Analysis*, DOI: 10.1007/978-3-642-19234-0_4,
© Springer-Verlag Berlin Heidelberg 2011

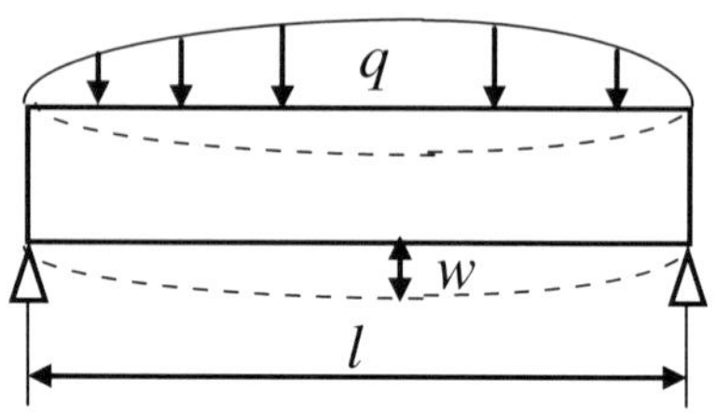

Fig. 4.1 A simply-supported beam

$$\begin{cases} \text{Equilibrium equation: } EI\dfrac{d^4w}{dx^4} = -q(x) \\[2ex] \text{Boundary conditions: } \quad x = 0 : w = \dfrac{d^2w}{dx^2} = 0 \\[2ex] \qquad\qquad\qquad\qquad x = l : w = \dfrac{d^2w}{dx^2} = 0 \end{cases} \qquad (4.1)$$

In (4.1), $l =$ beam length, $E = $ *Young's modulus*, and $I =$ cross-section moment and $EI=$ flexural rigidity of the beam with dimension FL^2 ($F =$ dimension of force). Distributed loading $q(x)$ can be expressed by characteristic parameters such as characteristic loading q_m and characteristic length l_q besides independent variable x.

Deflection of the beam is a function of parameters introduced in (4.1):

$$w = f(x,\ l;\ EI;\ q_m,\ l_q). \qquad (4.2)$$

This problem has two independent dimensions. Taking l and EI as a unit system produces:

$$\frac{w}{l} = f\left(\frac{x}{l},\ \frac{q_m l^3}{EI},\ \frac{l_q}{l}\right). \qquad (4.3)$$

If $\frac{q_m l^3}{EI}$ and $\frac{l_q}{l}$ are constants in the model and prototype, i.e.,

$$\left(\frac{q_m l^3}{EI}\right)_m = \left(\frac{q_m l^3}{EI}\right)_p \text{ and} \left(\frac{l_q}{l}\right)_m = \left(\frac{l_q}{l}\right)_p,$$

then dimensionless distribution of deflection is the same for model and prototype:

$$\frac{w}{l} = f\left(\frac{x}{l}\right). \qquad (4.4)$$

Modeling experiments do not require the geometry of the model to be similar to that of the prototype but do require the cross-section moment I to satisfy $\left(\frac{q_m l^3}{EI}\right)_m = \left(\frac{q_m l^3}{EI}\right)_p$. If material in the beam of the model is the same as material in the prototype, a square-shaped cross-section can simulate the *I*-shaped cross-section and a solid cross-section can be used to simulate a hollow one.

Fig. 4.2 Three-dimensional elastic body

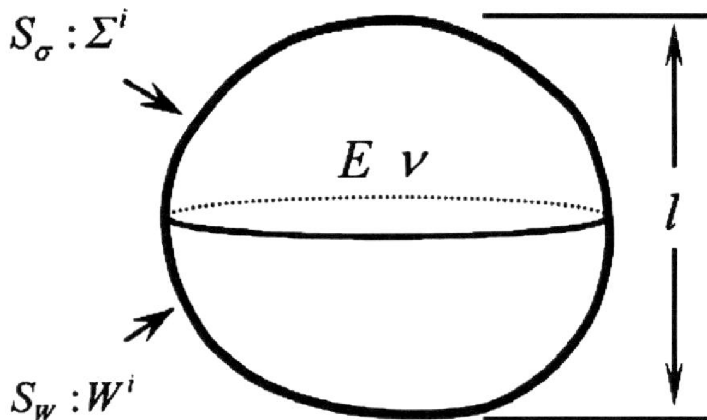

4.1.2 Deformation and Stress in Three-Dimensional Elastic Bodies

Under the action of applied forces, three-dimensional elastic bodies deform in both shape and volume (Fig. 4.2). Distribution of the induced stress and strain can be determined by the four governing factors:

Factor 1: Body geometry

Body geometry can be characterized by several lengths, e.g., l, l', l'' ..., where l can be selected as a characteristic length;

Factor 2: Material properties

For isotropic linearly elastic materials, material properties can be described by *Young's modulus E* and *Poisson's ratio v*;

Factor 3: Loading

Over a portion of body surface S_σ, distributed loading per unit surface area $\Sigma^i(x, y, z)$ is specified:

$$\Sigma^i = \Sigma \cdot f_\Sigma^i \left(\frac{x}{l}, \frac{y}{l}, \frac{z}{l} \right), \tag{4.5}$$

where Σ = characteristic distributed loading, f_Σ^i = dimensionless distribution function, and superscript "i" can be 1, 2, or 3, with each superscript identifying the component along a certain coordinate.

There can be additional N concentrated forces $F^i_{(n)}$ ($n = 1, 2, ..., N$) over the body surface. Dividing these forces $F^i_{(n)}$ by characteristic area l^2 converts them into distributed loading $\frac{F^i_{(n)}}{l^2}$ and their action points correspond to the distribution function of the distributed loading. Consequently, only distributed loading needs consideration. In general, body height h is so low that the order of magnitude of induced stress by gravitational force $\rho g h \ll$ characteristic distributed loading Σ or $\rho g h \ll$ critical loading stress p_{cr}, that makes a structure lose stability, and gravity effect can be ignored.

Factor 4: Constrained displacements

Assuming that constrained displacements W^i ($i = 1, 2, 3$) over another portion of body surface S_w are specified and rigid-body displacement is excluded;

$$W^i = W \cdot g^i \left(\frac{x}{l}, \frac{y}{l}, \frac{z}{l}\right), \tag{4.6}$$

where W = characteristic constrained displacement and g^i = dimensionless distribution function.

Based on these four governing factors, there should be four similarity requirements for modeling. Five corresponding reduction ratios correlate model with prototype:

1. Requirement for body geometry

Body geometry must be similar in model and prototype. Reduction ratio of length α_l:

$$\alpha_l = \frac{l_p}{l_m}. \tag{4.7}$$

2. Requirement for properties of material

Properties of material must be similar in model and prototype. Reduction ratio of Young's modulus $= \alpha_E$ and reduction ratio of Poisson's ratio $= \alpha_v$:

$$\alpha_E = \frac{E_p}{E_m} \quad \text{and} \quad \alpha_v = \frac{v_p}{v_m}. \tag{4.8}$$

3. Requirement for loadings

Distributed loading applied to model and prototype must be similar and applied at geometrically similar locations in model and prototype. Thus, when

$$\left(\frac{x^i}{l}\right)_p = \left(\frac{x^i}{l}\right)_m, \tag{4.9}$$

The distributed displacement should be:

$$f^i_{\Sigma p}\left(\left(\frac{x}{l}\right)_p, \left(\frac{y}{l}\right)_p, \left(\frac{z}{l}\right)_p\right) = f^i_{\Sigma m}\left(\left(\frac{x}{l}\right)_m, \left(\frac{y}{l}\right)_m, \left(\frac{z}{l}\right)_m\right). \tag{4.10}$$

Reduction ratio of characteristic distributed loading Σ is defined:

$$\alpha_\Sigma = \frac{\Sigma_p}{\Sigma_m} \quad \text{or} \quad \alpha_\Sigma = \frac{\left(F^i_{(n)}/l^2\right)_p}{\left(F^i_{(n)}/l^2\right)_m}, \tag{4.11}$$

where α_Σ is constant for both distributed loading and concentrated forces.

4. Requirement for constrained displacements

Constrained displacements must be similar and applied at geometrically similar locations in model and prototype, i.e.,:

$$g^i_p\left(\left(\frac{x}{l}\right)_p, \left(\frac{y}{l}\right)_p, \left(\frac{z}{l}\right)_p\right) = g^i_p\left(\left(\frac{x}{l}\right)_m, \left(\frac{y}{l}\right)_m, \left(\frac{z}{l}\right)_m\right). \tag{4.12}$$

Reduction ratio of characteristic constrained displacement W:

$$\alpha_W = \frac{W_p}{W_m}. \tag{4.13}$$

Requirements related to the four governing factors have five reduction ratios: α_l, α_E, α_v, α_Σ, and α_W. In addition, distributed loading $(f_{\Sigma p}^i = f_{\Sigma m}^i)$ and the same constrained displacements $(g_p^i = g_m^i)$ must be similarly applied at geometrically similar locations, $\left(\frac{x}{l}, \frac{y}{l}, \frac{z}{l}\right)_p = \left(\frac{x}{l}, \frac{y}{l}, \frac{z}{l}\right)_m$.

To summarize, displacements w^i, stresses σ_{ij}, and strains ε_{ij} in an elastic body can be expressed by independent variables representing four governing factors:

$$\begin{cases} w^i = f^i(l,\ l',\ l''\ldots;\ E,\ v;\ \Sigma,\ W;\ x,\ y,\ z), \\ \sigma_{ij} = g_{ij}(l,\ l',\ l''\ldots;\ E,\ v;\ \Sigma,\ W;\ x,\ y,\ z), \\ \varepsilon_{ij} = h_{ij}(l,\ l',\ l''\ldots;\ E,\ v;\ \Sigma,\ W;\ x,\ y,\ z). \end{cases} \tag{4.14}$$

There are two independent dimensions in the expressions in (4.14), so selecting fundamental quantities l and E as a unit system produces dimensionless function relationships:

$$\begin{cases} \dfrac{w^i}{l} = f^i\left(\dfrac{l'}{l},\ \dfrac{l''}{l}\ldots;\ v;\ \dfrac{\Sigma}{E},\ \dfrac{W}{l};\ \dfrac{x}{l},\ \dfrac{y}{l},\dfrac{z}{l}\right), \\[2mm] \dfrac{\sigma_{ij}}{E} = g_{ij}\left(\dfrac{l'}{l},\ \dfrac{l''}{l}\ldots;\ v;\ \dfrac{\Sigma}{E},\ \dfrac{W}{l};\ \dfrac{x}{l},\ \dfrac{y}{l},\dfrac{z}{l}\right), \\[2mm] \varepsilon_{ij} = h_{ij}\left(\dfrac{l'}{l},\ \dfrac{l''}{l}\ldots;\ v;\ \dfrac{\Sigma}{E},\ \dfrac{W}{l};\ \dfrac{x}{l},\ \dfrac{y}{l},\dfrac{z}{l}\right). \end{cases} \tag{4.15}$$

Because the problem possesses characteristics of linearity, effects of loadings and constrained displacements have additive property. Relationships (4.15) reduce to linear summation of both effects:

$$\begin{cases} \dfrac{w^i}{l} = \dfrac{\Sigma}{E} \cdot f_1^i\left(\dfrac{l'}{l},\ \dfrac{l''}{l}\ldots;\ v;\ \dfrac{x}{l},\ \dfrac{y}{l},\dfrac{z}{l}\right) \\[2mm] \qquad + \dfrac{W}{l} \cdot f_2^i\left(\dfrac{l'}{l},\ \dfrac{l''}{l}\ldots;\ v;\ \dfrac{x}{l},\ \dfrac{y}{l},\dfrac{z}{l}\right), \\[2mm] \dfrac{\sigma_{ij}}{E} = \dfrac{\Sigma}{E} \cdot g_{ij1}\left(\dfrac{l'}{l},\ \dfrac{l''}{l}\ldots;\ v;\ \dfrac{x}{l},\ \dfrac{y}{l},\dfrac{z}{l}\right) \\[2mm] \qquad + \dfrac{W}{l} \cdot g_{ij2}\left(\dfrac{l'}{l},\ \dfrac{l''}{l}\ldots;\ v;\ \dfrac{x}{l},\ \dfrac{y}{l},\dfrac{z}{l}\right), \\[2mm] \varepsilon_{ij} = \dfrac{\Sigma}{E} \cdot h_{ij1}\left(\dfrac{l'}{l},\ \dfrac{l''}{l}\ldots;\ v;\ \dfrac{x}{l},\ \dfrac{y}{l},\dfrac{z}{l}\right) \\[2mm] \qquad + \dfrac{W}{l} \cdot h_{ij2}\left(\dfrac{l'}{l},\ \dfrac{l''}{l}\ldots;\ v;\ \dfrac{x}{l},\ \dfrac{y}{l},\dfrac{z}{l}\right). \end{cases} \tag{4.16}$$

Concrete forms of $f_1^i, f_2^i, g_{ij1}, g_{ij2}, h_{ij1}$, and h_{ij2} can be determined by modeling experiments or theoretical analysis. According to *modeling law* (4.16), in addition to similarity requirements related to body geometry and distribution functions, dimensionless independent physical parameters $\frac{\Sigma}{E}, \frac{W}{l}$, and v must be identical for model and prototype:

$$\left(\frac{\Sigma}{E}, \frac{W}{l}, v \right)_m = \left(\frac{\Sigma}{E}, \frac{W}{l}, v \right)_p$$

Thus:

$$\frac{\Sigma_p}{\Sigma_m} = \frac{E_p}{E_m}, \quad \frac{W_p}{W_m} = \frac{l_p}{l_m}, \quad \text{and} \quad \frac{v_p}{v_m} = 1.$$

In other words, reduction ratios must satisfy:

$$\alpha_\Sigma = \alpha_E, \quad \alpha_W = \alpha_l \quad \text{and} \quad \alpha_v = 1.$$

Three of the reduction ratios $\alpha_\Sigma, \alpha_E, \alpha_W, \alpha_l$, and α_v are independent and in particular, $\alpha_v = 1$.

If material in the model and prototype is the same, i.e., $\alpha_E = 1$ and $\alpha_v = 1$, and if body geometry and distribution functions are identical for model and prototype, relative elongations, stresses, and strains are constant at similar locations and there is a *geometrically similar law*:

at

$$\left(\frac{x}{l}, \frac{y}{l}, \frac{z}{l} \right)_p = \left(\frac{x}{l}, \frac{y}{l}, \frac{z}{l} \right)_m :$$

$$\left(\frac{w^i}{l} \right)_p = \left(\frac{w^i}{l} \right)_m, \quad \left(\sigma_{ij} \right)_p = \left(\sigma_{ij} \right)_m, \quad \text{and} \quad \left(\varepsilon_{ij} \right)_p = \left(\varepsilon_{ij} \right)_m.$$

4.1.3 Application of Centrifugal Machine to Modeling Gravity Effect

In Sect. 4.1.2, gravity effect is ignored. However, *gravity effect* cannot be ignored if height of elastic body h exceeds a certain value and the order of magnitude of induced stress $\rho g h$ is close to characteristic distributed loading Σ, $\rho g h = O(\Sigma)$ or close to critical loading stress p_{cr}, that makes the structure lose stability, $\rho g h = O(p_{cr})$, where ρg is specific gravity of material. In such a case, specific gravity ρg becomes an additional independent variable in distribution functions of displacement, stress, and strain. For example, distribution of displacement:

$$w^i = f^i(l, \ l', \ l'' \ldots; \ E, \ v; \ \rho g; \ \Sigma, \ W; \ x, \ y, \ z). \tag{4.17}$$

Distributions of stress and strain can be dealt with similarly. Selecting l and E as a unit system produces dimensionless relationship:

$$\frac{w^i}{l} = f^i\left(\frac{l'}{l}, \ \frac{l''}{l} \ldots; \ v; \ \frac{\Sigma}{E}; \ \frac{\rho g l}{E}; \ \frac{W}{l}; \ \frac{x}{l}, \ \frac{y}{l}, \ \frac{z}{l}\right). \tag{4.18}$$

Since the problem possesses characteristics of linearity, effects of surface loading, volumetric loading (gravitational forces) and constrained displacement possess additive property and relationship (4.18) reduces:

$$\begin{aligned}
\frac{w^i}{l} = {} & \frac{\Sigma}{E} \cdot f_1^i\left(\frac{l'}{l}, \ \frac{l''}{l} \ldots; \ v; \ \frac{x}{l}, \ \frac{y}{l}, \ \frac{z}{l}\right) \\
& + \frac{\rho g l}{E} \cdot f_2^i\left(\frac{l'}{l}, \ \frac{l''}{l} \ldots; \ v; \ \frac{x}{l}, \ \frac{y}{l}, \ \frac{z}{l}\right) \\
& + \frac{W}{l} \cdot f_3^i\left(\frac{l'}{l}, \ \frac{l''}{l} \ldots; \ v; \ \frac{x}{l}, \ \frac{y}{l}, \ \frac{z}{l}\right).
\end{aligned} \tag{4.19}$$

In addition to requirements related to similarity of body geometry and identity of distribution functions, *modeling law* also requires:

$$\left(\frac{\Sigma}{E}, \ \frac{\rho g l}{E}, \ \frac{W}{l}, \ v\right) = \text{constant.}$$

Thus:

$$\frac{\Sigma_p}{\Sigma_m} = \frac{E_p}{E_m}, \ \frac{(\rho/E)_p}{(\rho/E)_m} \cdot \frac{g_p}{g_m} = \frac{l_p}{l_m}, \ \frac{W_p}{W_m} = \frac{l_p}{l_m}, \quad \text{and} \ \frac{v_p}{v_m} = 1.$$

If material in the model and prototype is the same, and model testing is carried out at nearly the same elevation, the second condition $\frac{(\rho/E)_p}{(\rho/E)_m} \cdot \frac{g_p}{g_m} = \frac{l_p}{l_m}$ reduces to the requirement for reduction ratio of length:

$$\alpha_l = \frac{l_p}{l_m} = 1.$$

Consequently, small-scale modeling is worthless.

To overcome difficulty in modeling *gravity effect*, Bucky (1931) originated the idea of a centrifugal machine in which adjustable centripetal acceleration simulates gravitational acceleration (Fig. 4.3). In that centrifugal machine, a belt driven by a motor rotates a central axle and a crossbeam fixed on that axle. Two hanging baskets are set up, one at each end of the crossbeam, with a testing model placed in one basket and a compatible weight placed in the other basket. To start the motor and control rotation velocity of axle and beam, centripetal force is applied to the model in the basket. If characteristic length of the model is sufficiently small and radius of rotation R (=1/2 of crossbeam length) is sufficiently large, centripetal

Fig. 4.3 Sketch of a centrifugal machine

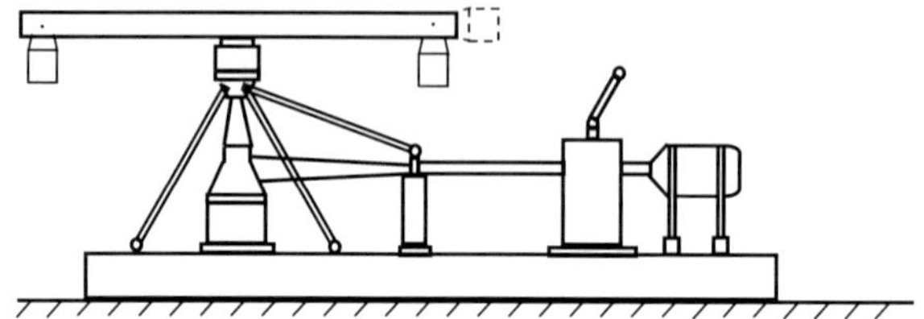

forces exerted on particles in the model are approximately parallel and point to the center of rotation. Because the model's linear velocity v is adjustable, proper centripetal acceleration $\frac{v^2}{R}$ simulates gravitational acceleration g.

To overcome difficulty in modeling gravity effect, centrifugal acceleration $\frac{v^2}{R}$ simulates gravitational acceleration g and corresponding similarity condition becomes:

$$\left(\frac{\rho(v^2/R)l}{E}\right)_m = \left(\frac{\rho g l}{E}\right)_p. \tag{4.20}$$

If material is the same as in model and prototype, condition (4.20) becomes:

$$\frac{v^2/R}{g} = \frac{l_p}{l_m} = \alpha_l. \tag{4.21}$$

In general, this condition can be satisfied without difficulty.

4.1.4 Photoelastic Experiments

Metals, concrete, and materials like rock and soil are opaque, making it difficult to inspect and measure deformation and stress without damaging what is being tested. In photoelastic experiments, models made of transparent photoelastic materials such as epoxy resin are set up in a polarized optical field. When a model is subjected to loading, an interferogram displays and records deformation distribution in the model. Stress distribution can be obtained through conversion from interference fringes.

The similarity law related to stress distribution derived in Sect. 4.1.2 converts:

$$\begin{aligned}
\frac{\sigma_{ij}}{E} = \frac{\Sigma}{E} \cdot g_{ij1}\left(\frac{l'}{l}, \frac{l''}{l}, \ldots; v; \frac{\Sigma}{E}, \frac{W}{l}; \frac{x}{l}, \frac{y}{l}, \frac{z}{l}\right) \\
+ \frac{W}{l} \cdot g_{ij2}\left(\frac{l'}{l}, \frac{l''}{l}, \ldots; v; \frac{\Sigma}{E}, \frac{W}{l}; \frac{x}{l}, \frac{y}{l}, \frac{z}{l}\right),
\end{aligned} \tag{4.22}$$

Similarity requires that model and prototype satisfy conditions for body material, body geometry, loadings and constrained displacements:

1. Material and loading:

$$\frac{E_m}{E_p} = \frac{\Sigma_m}{\Sigma_p} \quad \text{and} \quad v_m = v_p; \tag{4.23}$$

2. Geometry and constrained displacement:

$$\frac{l_m}{l_p} = \frac{W_m}{W_p}. \tag{4.24}$$

Similarity also requires model and prototype to have the same distribution functions for loading and constrained displacement. Through conversion, stresses in prototype at geometrically similar locations are:

$$\left(\sigma_{ij}\right)_p = \frac{E_p}{E_m} \cdot \left(\sigma_{ij}\right)_m. \tag{4.25}$$

It is difficult to satisfy requirements for *Poisson's ratio* $v_m = v_p$ in similarity conditions (4.23) because Poisson's ratio of transparent photoelastic material differs significantly from most prototype material. Using relationship (4.25) means that stress obtained by conversion is approximate, so error must be estimated and corrections to be made empirically.

For problems of plane stress or plane strain, advantages outweigh difficulties in satisfying requirement $v_m = v_p$ because the following two theorems are validated in the two-dimensional theory of elasticity:

Theorem 1 *If the region occupied by the body has simple connection, either with or without application of volumetric force and if, at the boundary, distributed loadings are the only constraint factor, then stress in the body does not relate to Poisson's ratio.*

Theorem 2 *If the region occupied by the body has complex connections, either without volumetric force or with constant volumetric forces being applied and if, at the boundary, distributed loadings are the only constraint factor and resultant force is zero, stress in the body does not relate to Poisson's ratio.*

For problems of plane stress or plane strain, if body geometries are similar and distributed loadings are the same in model and prototype, stress distribution satisfies similarity relationships:

$$\frac{\left(\sigma_{ij}\right)_p}{\left(\sigma_{ij}\right)_m} = \frac{\Sigma_p}{\Sigma_m} \quad \text{or} \quad \frac{\left(\sigma_{ij}\right)_p}{\left(\sigma_{ij}\right)_m} = \frac{\left(F_1/(hl)\right)_p}{\left(F_1/(hl)\right)_m,}$$

where $\frac{(F_1)_p}{(F_1)_m}$ = reduction ratio of external force and $\frac{h_p}{h_m}$ = reduction ratio of thickness of the two-dimensional body.

4.1.5 Critical Loading for Instability of Columns Under Compression

Engineering structures are systems that bear and transfer external loadings by combining elements such as beams, rods, axles, plates, and shells. Nowadays, there is wide use of structures such as thin rods, thin plates, and thin shells. In certain cases when such structures are under compression, even if inner compression stress is less than yield strength, there may be so much lateral bending that the structure loses bearing capacity. This phenomenon is called buckling instability.

A column of varying cross-section under compression with both ends fixed illustrates some general ideas about structural instability and cases having other boundary conditions can be treated similarly. If the column is subjected to gradually increasing axial force P, critical force $P = P_{cr}$ arises in which the column suddenly deforms into a bent state and buckles.

Assuming the length of the column of varying cross-section is l and axial compression force P is applied (Fig. 4.4) and assuming that the coordinate along the column length is x and distribution of cross-sectional moment of inertia is $I(x) = I_0 \cdot I^*\left(\frac{x}{l}\right)$, where $I_0 = $ a characteristic moment (with dimension L^4), and $I^* = $ distribution function.

Mechanics of materials dictate that lateral displacement of straight column w satisfies equation and boundary condition:

$$\begin{cases} \text{Equation: } E\dfrac{d^2\left[I(x)d^2w/dx^2\right]}{dx^2} + P\dfrac{d^2w}{dx^2} = 0 \\[2mm] \text{or} \quad \dfrac{EI_0}{P}\dfrac{d^2\left[I^*(x)d^2w/dx^2\right]}{dx^2} + \dfrac{d^2w}{dx^2} = 0; \\[2mm] \text{Boundary condition:} \begin{cases} x = 0: \quad w = \dfrac{dw}{dx} = 0, \\[2mm] x = l: \quad w = \dfrac{dw}{dx} = 0. \end{cases} \end{cases} \tag{4.26}$$

The problem in (4.26) has a trivial solution $w = 0$. Nontrivial solution functions $w(x)$ are generally for special constant coefficient $\frac{EI_0}{P}$ only when distribution function $= I^*\left(\frac{x}{l}\right)$. The $w(x)$ termed eigenfunction belongs to eigenvalue $\frac{EI_0}{P}$. In fact, the nontrivial solution is controlled by governing parameters $\frac{EI_0}{P}$, l, and distribution function $I^*\left(\frac{x}{l}\right)$. If material is specified, critical loading P_{cr} satisfies:

$$\frac{P_{cr}}{EI_0} = f(l), \tag{4.27}$$

Fig. 4.4 Column under
compression

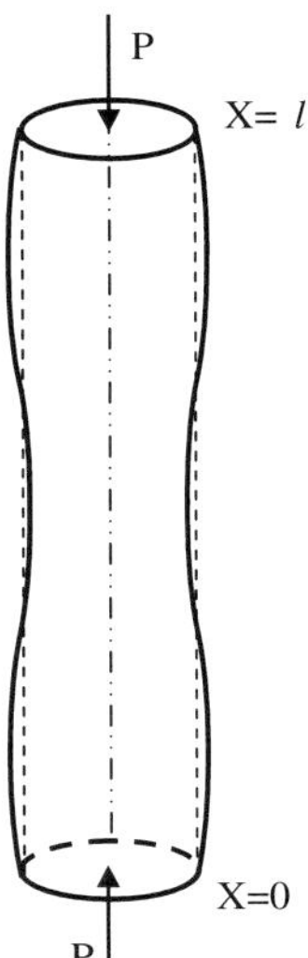

where the concrete form of function f on the right hand side depends on distribution function $I^*\left(\frac{x}{l}\right)$.

In relationship (4.27), the dimension of the left hand side is L^{-2} and the dimension of the independent variable of function f is L. Therefore, only one independent dimension in this problem produces dimensionless relationship:

$$\frac{P_{cr}l^2}{EI_0} = \text{constant}, \tag{4.28}$$

where the constant on the right hand side is related to distribution function $I^*\left(\frac{x}{l}\right)$.

Two conclusions are obtainable:

1. For the given distribution function of cross-section, critical loading P_{cr} is proportional to $\frac{EI_0}{l^2}$.
2. In modeling experiments, the shape of cross-sections does not need to be geometrically similar.It is enough if the model and prototype have the same distribution function $I^*\left(\frac{x}{l}\right)$

4.2 Vibration and Wave Motion of Elastic Bodies

Dimensional analysis is applicable to analyzing inherent vibration and forced vibration as well as analyzing wave propagation in elastic bodies.

4.2.1 Inherent Vibration of Finite Elastic Bodies

For given boundary conditions, a finite elastic body has inherent frequency. Vibration in elastic bodies is essentially a kind of energy transformation between kinetic and potential energies. Kinetic energy originates from medium *inertia*, represented by ρ. Elastic energy originates from deformation recoverability of the elastic medium characterized by *Young's modulus E* and *Poisson's ratio v*. If given boundary conditions are displacement fixed conditions and characteristic length of elastic body $= l$, inherent frequency of the elastic body ω is:

$$\omega = f(l,\ l',\ l'',\ \ldots,\ \rho,\ E,\ v). \tag{4.29}$$

Taking $l,\ \rho,\ E$ as a unit system produces the dimensionless relationship:

$$\frac{\omega l}{\sqrt{E/\rho}} = f\left(\frac{l'}{l},\ \frac{l''}{l},\ \ldots,\ v\right). \tag{4.30}$$

For geometrically similar elastic bodies, inherent frequency is inversely proportional to characteristic length l and inherent period is proportional to characteristic length l. If model material is the same as prototype material, relationship simplifies:

$$\frac{\omega_p}{\omega_m} = \frac{l_m}{l_p}. \tag{4.31}$$

In other words, the reduction ratio of inherent frequency is the inverse of the reduction ratio of characteristic length: $\alpha_\omega = \frac{1}{\alpha_l}$. Cases involving other boundary conditions can be treated similarly.

4.2.2 Forced Vibration of Elastic Bodies

Unlike inherent vibration, forced vibration is subject to external loadings. Parameters governing forced vibration include:

1. Geometry of body $l,\ l',\ l'',\ \ldots$;
2. Material properties $\rho,\ E,\ v$;
3. Distributed loadings $\Sigma \cdot f_\Sigma^i\left(\frac{x}{l},\ \frac{y}{l},\ \frac{z}{l},\ \frac{t}{t_*}\right)$ applied on surface S_σ,
4. Concentrated loadingsconverted into distributed loadings (see Factor 3 in Sect. 4.1.2)
5. Constrained displacements $W \cdot f_W^i\left(\frac{x}{l},\ \frac{y}{l},\ \frac{z}{l},\ \frac{t}{t_*}\right)$ applied to surface S_W,

where $\Sigma =$ characteristic loading, $W =$ characteristic displacement, $f_\Sigma^i =$ distribution functions of distributed loadings, $f_W^i =$ distribution functions of constrained displacements, and t_* in the distribution function of distributed loadings is a characteristic inherent period of the elastic body $\frac{l}{\sqrt{E/\rho}}$.

Stress distribution in an elastic body can be obtained in a way similar to static analysis in Sect. 4.1.2:

$$\sigma_{ij} = g_{ij}(l,\ l',\ldots;\ \rho,\ E,\ v;\ \Sigma,\ W;\ x,\ y,\ z,\ t), \tag{4.32}$$

where function g_{ij} also depends on distribution functions of loadings and constrained displacements.

Taking $l,\ \rho,\ E$ as a unit system produces dimensionless stresses:

$$\begin{aligned}
\frac{\sigma_{ij}}{E} = {} & \frac{\Sigma}{E} \cdot p_{ij}\left(\frac{l'}{l},\ldots;\ v;\ \frac{x}{l},\ \frac{y}{l},\ \frac{z}{l},\ \frac{t}{l/\sqrt{E/\rho}}\right) \\
& + \frac{W}{l} \cdot q_{ij}\left(\frac{l'}{l},\ldots;\ v;\ \frac{x}{l},\ \frac{y}{l},\ \frac{z}{l},\ \frac{t}{l/\sqrt{E/\rho}}\right)
\end{aligned} \tag{4.33}$$

Small-scale model testing requires that conditions are satisfied:

1. Geometry of the model and prototype are similar $\alpha_l = \frac{l_p}{l_m}$;
2. Poisson's ratios of the model material and prototype material are identical: $\alpha_v = 1$;
    Reduction ratios of Young's modulus and characteristic distributed loadings are identical: $\alpha_E = \alpha_\Sigma$;
3. Constrained displacements of the model and prototype are geometrically similar: $\alpha_W = \alpha_l$.
    Thus, at similar locations and similar moments

$$\left(\frac{x}{l},\ \frac{y}{l},\ \frac{z}{l},\ \frac{t}{l/\sqrt{E/\rho}}\right)_m = \left(\frac{x}{l},\ \frac{y}{l},\ \frac{z}{l},\ \frac{t}{l/\sqrt{E/\rho}}\right)_p :$$

$$\left(\frac{\sigma_{ij}}{E}\right)_p = \left(\frac{\sigma_{ij}}{E}\right)_m \quad \text{or} \quad \frac{(\sigma_{ij})_p}{(\sigma_{ij})_m} = \frac{E_p}{E_m}.$$

4.2.3 Wave Velocity of Body Waves and Surface Waves in Elastic Bodies and Wave Dispersion in Elastic Waveguides

Waves in elastic bodies can be classified as body waves and surface waves. The wave source of body waves is in an elastic body and an induced disturbance propagates in a finite distance and affects a finite region for a finite time. The wave source of surface waves is on an elastic surface and disturbances induced propagate along the surface and affect a finite surface region for a finite time.

For linearly elastic media, propagation of elastic waves possesses characteristics of linearity. An elastic wave can decompose as a linear combination of a series of harmonic waves. Dimensional analysis can be applied to the propagation velocity of harmonic waves, especially of body waves, surface waves, and waves in elastic waveguides.

1. Wave velocity of body waves

Assuming that the wave length of a harmonic wave is λ and that the property of inertia of elastic medium is density ρ, property of elastic recoverability is *Young's modulus* E and Poisson's ratio v. Wave velocity of a body wave is a function of governing parameters ρ, E, v, and λ:

$$c = f(\rho,\ E,\ v;\ \lambda). \tag{4.34}$$

Taking ρ, E, λ as a unit system produces dimensionless relationship:

$$\frac{c}{\sqrt{E/\rho}} = f(v) \quad \text{or } c = \sqrt{\frac{E}{\rho}} \cdot f(v). \tag{4.35}$$

This dimensionless relationship shows that wave velocity does not relate to wave length, so elastic body waves are non-dispersive. Furthermore, wave velocity is proportional to $\sqrt{\frac{E}{\rho}}$ and proportion coefficient $f(v)$ is a function of Poisson's ratio.

The theory of elastic dynamics [1, pp. 57–58] classifies elastic body waves into longitudinal and transversal waves. Longitudinal wave velocity c_d:

$$c_d = \sqrt{\frac{1 - v}{(1 + v)(1 - 2v)}} \cdot \sqrt{\frac{E}{\rho}}; \tag{4.36a}$$

Transversal wave velocity c_s:

$$c_s = \sqrt{\frac{1}{2(1 + v)}} \cdot \sqrt{\frac{E}{\rho}}. \tag{4.36b}$$

The corresponding proportion coefficient $f(v)$ of the longitudinal wave is $\sqrt{\frac{1-v}{(1+v)(1-2v)}}$. The corresponding proportion coefficient of the transversal wave is $\sqrt{\frac{1}{2(1+v)}}$.

2. Wave velocity of surface waves

Parameters governing a surface wave (known as the *Rayleigh wave* in honor of the first person to analyze wave nature) are the same as parameters that govern body waves. The difference between a surface wave and a body wave is that the source of surface waves is at the surface while the source of body waves is in the body. Similar to the discussion of wave velocity of body waves, the wave velocity of the *Rayleigh wave* c_R has the same formal expression as that of a body wave:

$$\frac{c_R}{\sqrt{E/\rho}} = f_R(v), \quad \text{or} \quad c_R = \sqrt{E/\rho} \cdot f_R(v). \tag{4.37}$$

However, the concrete form of proportion coefficient $f_R(v)$ differs from that of body waves $f(v)$.

The theory of elastic dynamics [1, p.147] shows that c_R is a solution for the algebraic equation:

$$\left(\frac{c_R}{c_s}\right)^6 - 8\left(\frac{c_R}{c_s}\right)^4 + \left(24 - \frac{16}{k^2}\right)\left(\frac{c_R}{c_s}\right)^2 - 16\left(1 - \frac{1}{k^2}\right) = 0, \tag{4.38}$$

where c_s = wave velocity of transverse waves and k^2 = square of the ratio of wave velocity for a longitudinal wave to wave velocity for a transversal wave:

$$k^2 = \frac{2(1-v)}{1-2v} > 1.$$

Wave velocity of the *Rayleigh wave* c_R is proportional to $\sqrt{\frac{E}{\rho}}$ and the proportion coefficient is a function of Poisson's ratio. Furthermore, the Rayleigh wave is a non-dispersive wave. Clearly, the variable $\left(\frac{c_R}{c_s}\right)^2$ in the algebraic equation (4.38) has a real root between 0 and 1:

$$0 < \left(\frac{c_R}{c_s}\right)^2 < 1. \tag{4.39}$$

Therefore, wave velocity of the Rayleigh wave is less than that of both body waves:

$$0 < c_R < c_s < c_d. \tag{4.40}$$

3. Wave velocity in wave guides

Elastic wave guides refer to elastic bodies having more or less parallel boundaries that guide elastic wave propagation along the direction parallel to those boundaries. Rods, plates, shells and layered elastic solids are examples of waveguides.

Using a cylinder with a free surface as an example, it is possible to suppose a compression (or tension) force being applied to an end of the cylinder that induces a disturbance that propagates along the length of the cylinder. Parameters that govern elastic wave velocity in the cylinder except those governing wave velocities of body waves or surface waves include the radius of the cylinder R as an additional parameter geometrically characterizing the cylindrical wave guide. Thus, wave velocity c of a harmonic wave in an elastic cylinder:

$$c = f(\rho; \, E, \, v; \, \lambda, \, R). \tag{4.41}$$

Taking ρ, E, and R as a unit system produces dimensionless relationship:

$$\frac{c}{\sqrt{E/\rho}} = f(v; \frac{\lambda}{R}).$$ (4.42)

Function f also depends on free boundary conditions at the cylinder surface. Relationship (4.42) shows that wave velocity depends on material properties and on wave length λ, meaning that the wave in the cylinder is dispersive. Dimensionless parameter $\frac{\lambda}{R}$ plays a distinctive role, so the cylinder radius as a characteristic length of the cylindrical wave guide is a key factor. This conclusion holds for other types of wave guides, such as plates and shells in which characteristic length is plate thickness or shell thickness.

Pochhammer's classic study on wave propagation in elastic cylinders (1876) [2] shows that wave velocity of harmonic compression waves c satisfies:

$$\left[\left(\frac{c}{c_s}\right)^2 + 2\right] \cdot \frac{\xi_d J_0(\xi_d)}{J_1(\xi_d)} + 4\left[\left(\frac{c}{c_d}\right)^2 - 1\right] \cdot \frac{\xi_s J_0(\xi_s)}{J_1(\xi_s)} = 2\left(\frac{c}{c_s}\right)^2 \cdot \left[\left(\frac{c}{c_d}\right)^2 - 1\right],$$ (4.43)

where c_d = longitudinal wave velocity, c_s = transversal wave velocity, ξ_d and ξ_s are expressed by relative wave velocity $\frac{c}{c_d}$, relative wave velocity $\frac{c}{c_s}$, and relative wave length $\frac{\lambda}{R}$:

$$\xi_d = \frac{2\pi R}{\lambda}\sqrt{\left(\frac{c}{c_d}\right)^2 - 1} \quad \text{and} \quad \xi_s = \frac{2\pi R}{\lambda}\sqrt{\left(\frac{c}{c_s}\right)^2 - 1},$$ (4.44)

where J_0 = ordinary Bessel functions of the first kind of order zero and J_1 = ordinary Bessel functions of the first kind of order 1.

Wave velocity c in a cylinder is proportional to $\sqrt{\frac{E}{\rho}}$, except that the form of proportion coefficient $f(v; \frac{\lambda}{R})$ is relatively complicated.

For long waves in a slender cylinder, $\frac{R}{\lambda} \ll 1$, compression wave velocity can be approximated:

$$\frac{c}{\sqrt{E/\rho}} = 1 - \pi^2 v^2 \left(\frac{R}{\lambda}\right)^2.$$ (4.45)

In this case, the concrete form of $f(v; \frac{\lambda}{R})$ is relatively simple:

$$f(v, \frac{\lambda}{R}) = 1 - \pi^2 v^2 \left(\frac{R}{\lambda}\right)^2.$$

If $\frac{R}{\lambda} \ll \frac{1}{\pi v}$, there is an approximate expression that is commonly used for compression wave velocity in slender cylinders:

$$c = \sqrt{\frac{E}{\rho}}. \tag{4.46}$$

Only in such a case, compression wave in cylinders is non-dispersive.

4.3 Stress Analysis of Elasto-Plastic Bodies

Engineering problems often include elastic deformation and nonlinear plastic deformation. However, there is still a lack of understanding of the physical substance and a lack of mathematical description of plastic deformation. In comparison with elastic deformation, the complexity of plastic deformation lies in the dual nonlinearity involved in the constitutive relationship of materials, i.e., non-linearity during *plastic loading* and subsequent non-linear hysteresis loops in the unloading–reloading process.

No reliable theory describes mechanical behavior of materials that take arbitrary and complicated paths. However, with proper assumptions about elasto-plastic behavior, modeling engineering problems can still produce useful conclusions. Considering *static tension of cylindrical rods* and discussing modeling of stress–strain relationship and stress distribution in elasto-plastic bodies, allows analysis of cold-rolling steel plates in pressure working of metallic materials. Besides, dimensional analysis of the physical substance of hardness can be introduced.

4.3.1 Static Tension of Cylindrical Rods

Figure 4.5 indicates the stress–strain curve of a *cylindrical rod under simple tension*. When tensile stress increases gradually from zero, stress σ and strain ε increase in direct proportion along a straight line with a slope equal to *Young's modulus E* until yield point $(\varepsilon_s, \sigma_s)$, where σ_s = yield limit and ε_s = corresponding strain. If stress increases further, irreversible plastic deformation begins in which stress and strain no longer increase along the linear elastic section and begin to follow a curve until stress reaches maximum σ_b, called tensile strength. Corresponding strain is denoted by ε_b. Thereafter, cylindrical rod material cannot bear more stress and stress begins to decrease. Meanwhile, strain continues to increase until fracture point $(\varepsilon_m, \sigma_m)$, where ε_m = maximum strain and σ_m = corresponding stress. Fig. 4.5a shows stress–strain relationship in the loading process and Fig. 4.5b shows corresponding dimensionless relationship in which σ_s is used as a unit of stress. If two materials can have the same dimensionless stress–strain relationships, it is proper to simulate loading, stress and strain necking in tensile behavior of cylindrical rods made of those two materials.

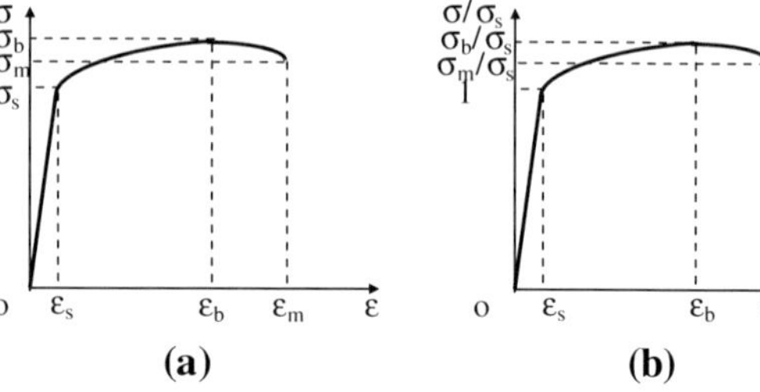

Fig. 4.5 $\sigma \sim \varepsilon$ curve for simple tension

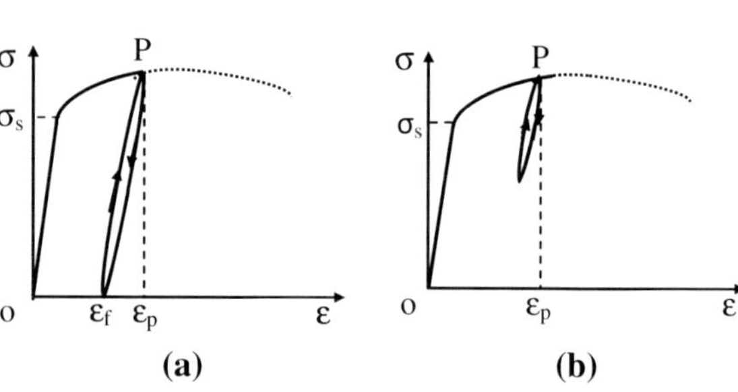

Fig. 4.6 Unloading and reloading after plastic loading

The case under discussion concerns non-linearity of plastic tension in simple loading. If an unloading and reloading process occurs after plastic deformation of simple loading, stress–strain relationship becomes more complicated because unloading and reloading are both non-linear and there is a non-linear hysteresis loop (Fig. 4.6). Figure 4.6a indicates that unloading starts from plastic state point P and ends at 0 stress point. The unloading curve moves away from the original loading section and descends suddenly along an unloading path to point $\left(\varepsilon_f, \, 0\right)$, where ε_f = residual strain in the material. If a new reloading process starts from point $\left(\varepsilon_f, \, 0\right)$, states of stress and strain follow a new loading path until reaching starting point P of previous unloading. There is completion of a hysteresis loop that is composed of the unloading curve section and the reloading section. Work done by external loading in the hysteresis loop converts irreversibly into energies, such as heat and dislocation. If loading continues when the reloading state returns to point P, the loading path follows the extension of the earlier loading path (dashed line in Fig. 4.6). If unloading that starts at loading plastic point P does not reach 0 stress point $\left(\varepsilon_f, \, 0\right)$ and reloading occurs, a small non-linear hysteresis loop occurs (see Fig. 4.6b).

To achieve the modeling objective, requires that modeling material and prototype material have the same dimensionless stress–strain relationships for arbitrary loading and unloading processes.

Examining shear deformation requires dimensionless shear stress–shear strain relationship that is similar to the case involving tension. In reality, loading and unloading paths in a structure differ from point to point and, strictly speaking, model material and prototype material must have the same dimensionless stress–strain relationships during arbitrary loading and unloading. It is difficult to satisfy such requirements in model material. Therefore, in modeling elasto-plastic deformation problems, material used is generally the same in model and prototype, and although model size changes, model and prototype are geometrically similar.

To model a body undergoing elasto-plastic deformations, procedure and formulation are similar to the treatment of deformation and stress in three-dimensional elastic bodies Sect. 4.1.2. However, relevant parameters of plasticity must appear in the list of governing parameters. As an approximation, the paths of the unloading and reloading follow a straight line with a slope equal to Young's modulus E. Thus, governing parameters that describe constitutive relationship for the materials are not only elastic, i.e., Young's modulus and Poisson's ratio, but also plastic, such as σ_s, σ_b, ε_b, σ_m, and ε_m.

4.3.2 Modeling Stress Distribution in Elasto-Plastic Bodies

If external loadings and constrained displacements applied to the bodies are expressed by corresponding characteristic loading and displacement as well as by corresponding distribution functions, stress distribution in bodies can be:

$$\sigma_{ij} = g_{ij}(l;\ E,\ v,\ \sigma_s,\ \sigma_b,\ \varepsilon_b,\ \sigma_m,\ \varepsilon_m;\ \Sigma,\ W;\ x,\ y,\ z), \qquad (4.47)$$

where l = characteristic length of body geometry, Σ = characteristic loading and W = characteristic displacement.

Taking l and σ_s as a unit system, produce dimensionless stress distribution:

$$\frac{\sigma_{ij}}{\sigma_s} = g_{ij}\left(\frac{E}{\sigma_s},\ v,\ \frac{\sigma_b}{\sigma_s},\ \varepsilon_b,\ \frac{\sigma_m}{\sigma_s},\ \varepsilon_m;\ \frac{\Sigma}{\sigma_s},\ \frac{W}{l};\ \frac{x}{l},\ \frac{y}{l},\ \frac{z}{l}\right). \qquad (4.48)$$

Using prototype materials as model materials, is the easiest way to obtain perfect simulation results, allowing relationship (4.48) to simplify:

$$\frac{\sigma_{ij}}{\sigma_s} = g_{ij}\left(\frac{\Sigma}{\sigma_s},\ \frac{W}{l};\ \frac{x}{l},\ \frac{y}{l},\ \frac{z}{l}\right). \qquad (4.49)$$

If:

1. Body geometries are similar,
2. Constrained displacements are similar (i.e., $\alpha_W = \alpha_l$) and corresponding distribution functions are the same,
3. Loadings are similar (i.e., $\alpha_\Sigma = \frac{\Sigma_p}{\Sigma_m} = 1$) and corresponding distribution functions are the same,

then the same stresses occur at geometrically similar locations, i.e.,

$$\text{at } \left(\frac{x}{l},\ \frac{y}{l},\ \frac{z}{l}\right)_p = \left(\frac{x}{l},\ \frac{y}{l},\ \frac{z}{l}\right)_m :$$

$$\left(\frac{\sigma_{ij}}{\sigma_s}\right)_p = \left(\frac{\sigma_{ij}}{\sigma_s}\right)_m, \quad \text{or } (\sigma_{ij})_p = (\sigma_{ij})_m.$$

If material in the model and prototype differs, modeling requires identical dimensionless stress–strain relationships in model and prototype material.

Similarly, in stability analysis of structures, the conversion formula of critical loading for model F_m to critical loading for prototype F_p:

$$\frac{F_p}{l_p^2} = \frac{F_m}{l_m^2} \cdot \frac{(\sigma_s)_p}{(\sigma_s)_m} \tag{4.50}$$

4.3.3 Modeling Cold-Rolled Steel Plates

Figure 4.7 is a schematic diagram of the profile of cold-rolling of a steel plate. Assuming plate breadth $= b$, thickness before rolling $= l$, thickness after rolling $= l'$, roller radius $= R$, rotation velocity $= \Omega$; assuming that the steel plate possesses material properties: density $= \rho$, elastic constants $= $ *Young's modulus* E and Poisson's ratio v, plastic parameters $= \sigma_s$, σ_b, ε_b, σ_m, and ε_m, assuming that the stress–strain relationship is insensitive to strain rate, and assuming that no constraint is applied to plate extremities, that roller and rolling mill are rigid and that no sliding occurs between plate and roller, force exerted on mill F:

$$F = f(b,\, l,\, l',\, R,\, \Omega,\, \rho,\, E,\, v,\, \sigma_s,\, \ldots). \tag{4.51}$$

Taking l, ρ, and σ_s as a unit system, generates dimensionless relationship:

$$\frac{F}{\sigma_s l^2} = f\left(\frac{b}{l},\, \frac{l'}{l},\, \frac{R}{l};\, \frac{\Omega R}{\sqrt{\sigma_s/\rho}};\, \frac{E}{\sigma_s},\, v,\, \ldots\right) \tag{4.52a}$$

or

$$\frac{F}{\sigma_s b R} = f\left(\frac{b}{l},\, \frac{l'}{l},\, \frac{R}{l};\, \frac{\Omega R}{\sqrt{\sigma_s/\rho}};\, \frac{E}{\sigma_s},\, v,\, \ldots\right), \tag{4.52b}$$

where the square of the fourth similarity criterion number on the right hand side, $\frac{\rho\Omega^2 R^2}{\sigma_s}$, characterizes ratio of inertia effect to *strength effect*. In general, $\frac{\rho\Omega^2 R^2}{\sigma_s} \ll 1$ and can be ignored. The first three independent dimensional variables are similarity criterion numbers in geometry and the other two variables characterize

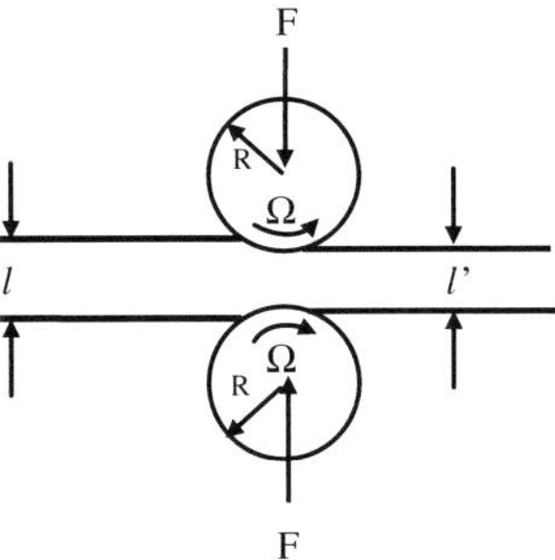

Fig. 4.7 Sketch of cold-rolling of steel plate

material properties of elasto-plasticity. If material is the same in the model and prototype:

$$\left(\frac{E}{\sigma_s}, v, \ \ldots\right) = \text{constant},$$

and if model geometry and prototype geometry are similar:

$$\left(\frac{b}{l}, \frac{l'}{l}, \frac{R}{l}\right) = \text{constant},$$

there is a similarity law for force applied to the rolling mill:

$$\frac{F}{\sigma_s bR} = \text{constant} \quad \text{or} \quad \frac{F}{bR} = \text{constant}.$$

Therefore, reduction ratio of force is:

$$\frac{F_p}{F_m} = \frac{(bR)_p}{(bR)_m} = \alpha_l^2.$$

In reality, strain rate $\frac{d\varepsilon}{dt}$ generated in the rolling process is not a negligible quantity in comparison with the strain rate $\left(\frac{d\varepsilon}{dt}\right)_0$ used in material testing and the main effect of the strain rate is to raise yield strength σ_s. Correction is possible by using corresponding yield strength σ_s under the mean strain rate in the rolling process. After such correction, above discussions and conclusions remain valid.

In pressure working of metallic materials, plastic deformation $\gg$ elastic deformation. Therefore, material may be assumed to be perfectly plastic or material may be assumed to follow the stress–strain relationship of power-law type $\sigma = \sigma_0 \varepsilon^n$. In such cases, requirements can be relaxed and a wider range of materials can be selected for modeling.

4.3.4 Hardness

For nearly one hundred years, indentation experiments have been performed for measuring hardness of materials. Formerly, hardness was used to estimate the strength index of materials, based on the assumption that yield strength is proportional to hardness and that proportional coefficient is an empirical constant. In recent years, there has been increased interest in indentation because of significantly improved indentation equipment and the need for measuring mechanical properties of materials on small scales. *Cheng* and *Cheng* [3] applied dimensional analysis and *finite element calculations* to reveal the physical substance involved in indentation depth and the relationship between indentation behavior and mechanical properties of materials.

Assuming a rigid conical indenter of given semi-angle θ (e.g., 68°) that indents normally into elasto-plastic solid material, friction at the contact surface is assumed to be negligible. In this case, hardness H is defined by

$$H = \frac{F}{\pi a^2},\tag{4.53}$$

where F is load and a is contact radius. Fig. 4.8 shows a conical indentation that concerns load F and depth for contact part h_c, from which contact radius a and hardness under load H can be evaluated:

$$a = h_c \tan \theta,\tag{4.54}$$

and

$$H = \frac{F}{\pi a^2}$$

Stress–strain curves of solids under uniaxial tension are assumed to be:

$$\left\{\begin{array}{ll} \sigma = E\varepsilon & \text{for } \varepsilon \le \dfrac{Y}{E}, \\[2mm] \sigma = K\varepsilon^n & \text{for } \varepsilon \ge \dfrac{Y}{E}, \end{array}\right.\tag{4.55}$$

where E = *Young's modulus*, Y = yield limit, K = strength coefficient, and n = work-hardening exponent. To satisfy the requirement of continuity, K must equal $Y\left(\frac{E}{Y}\right)^n$. When n = zero, material becomes elastic-perfect plastic. For most metallic materials, n has a value between 0.1 and 0.5.

As material properties are given, relationship between load F and indenter displacement h can be calculated by finite element methods. Dimensional analysis can apply to loading stage and unloading stage in indentation.

1. Dimensional analysis applied to loading stage

For given material and indenter with semi-angle θ, load F, and contact depth h_c must be uniquely determined by indenter displacement h:

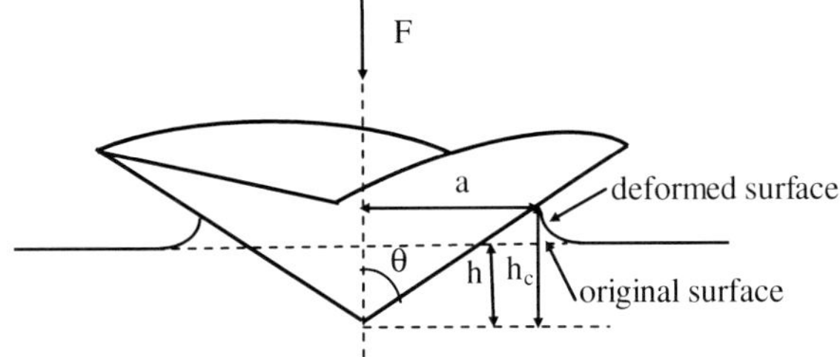

Fig. 4.8 Sketch of conical indentation

$$F = f_F(E, \, v; \, Y, \, n; \, h, \, \theta), \tag{4.56}$$

and

$$h_c = f_{h_c}(E, \, v; \, Y, \, n; \, h, \, \theta), \tag{4.57}$$

where $v = $ Poisson's ratio.

Because hardness is defined as:

$$H = \frac{F}{\pi a^2}$$

while

$$a = h_c \tan \theta,$$

H is also a function of the same independent variables as those of h_c:

$$H = f_H(E, \, v; \, Y, \, n; \, h, \, \theta). \tag{4.58}$$

Two independent dimensions appear among parameters E, v, Y, n, h, and θ. Taking E and h as a unit system produces:

$$\frac{F}{Eh^2} = f_F\left(\frac{Y}{E}, \, v, \, n, \, \theta\right), \tag{4.59}$$

$$\frac{h_c}{h} = f_{h_c}\left(\frac{Y}{E}, \, v, \, n, \, \theta\right), \tag{4.60}$$

$$\frac{H}{E} = f_H\left(\frac{Y}{E}, \, v, \, n, \, \theta\right). \tag{4.61}$$

Relative hardness $\frac{H}{E}$ converts:

$$\frac{H}{Y} = g_H\left(\frac{Y}{E}, \, v, \, n, \, \theta\right). \tag{4.62}$$

Fig. 4.9 $F/(Eh^2)$ versus Y/E and n

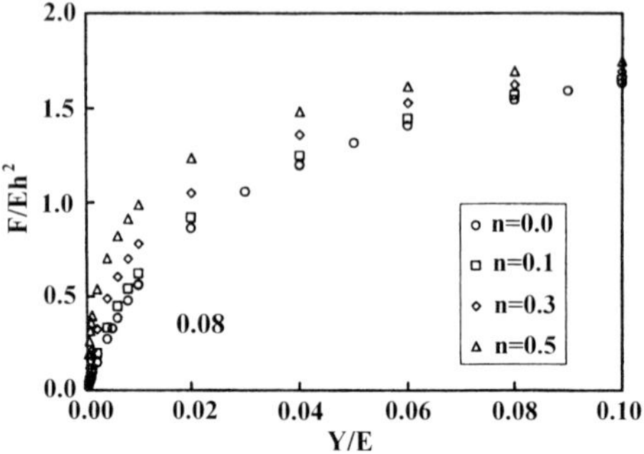

Fig. 4.10 H/Y versus Y/E and n

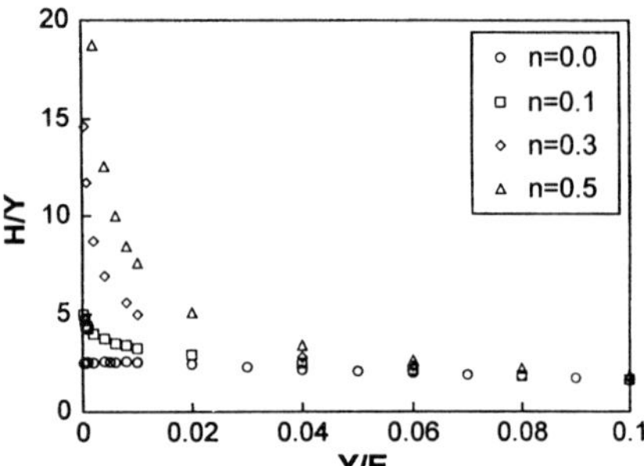

Figures 4.9 and 4.10 show finite element calculations of $\frac{F}{Eh^2}$ and $\frac{H}{Y}$ with variation of $\frac{Y}{E}$ and n for the given semi-angle of the indenter ($\theta = 68^\circ$) and given Poisson's ratio ($v = 0.3$).

Figure 4.10 shows that $\frac{H}{Y}$ is not simply a constant. H is approximately proportional to Y only when $n = 0$ or $\frac{Y}{E} > 0.04$, but the latter is unrealistic because $\frac{Y}{E} \ll 0.04$ in most materials. In general, $\frac{H}{Y}$ depends on $\frac{Y}{E}$, n and v.

2. Dimensional analysis applied to unloading stage

Load F in the stage of loading, satisfies:

$$\frac{F}{Eh^2} = f_F(\frac{Y}{E}, \ v, \ n, \ \theta).$$

When load causes elastic and plastic deformation in material and the indenter reaches maximum depth h_m, total work done by indenter W_{tot}:

$$W_{tot} = \int_0^{h_m} F\,dh = \frac{Eh_m^3}{3} \cdot f_F\left(\frac{Y}{E}, \ v, \ n, \ \theta\right). \tag{4.63}$$

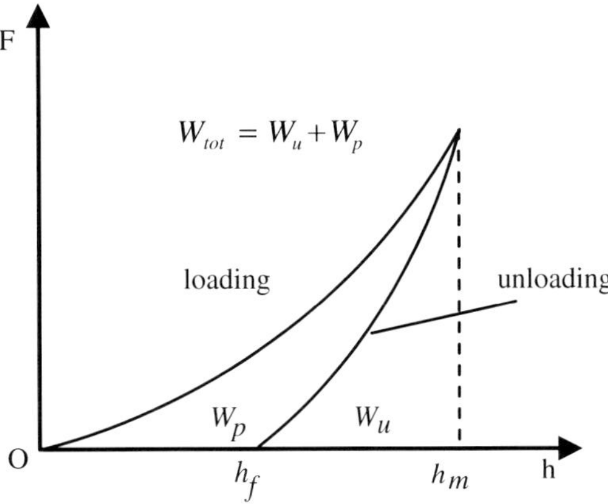

Fig. 4.11 Work done during loading and unloading

Thus, force F is proportional to h_m^2 and work W_{tot} is proportional to h_m^3.

Unloading starts after the indenter reaches maximum depth h_m. Thereafter, state points during unloading follow a different path from loading one (Fig. 4.11). Obviously, unloading path depends on h_m and force F during unloading can be:

$$F = Eh^2 \cdot f_u\left(\frac{Y}{E}, \frac{h}{h_m}, v, n, \theta\right). \tag{4.64}$$

In general, force F is no longer simply proportional to the square of indenter displacement h, but also depends on ratio $\frac{h}{h_m}$, which is an independent variable of dimensionless function f_u.

When force F first becomes 0 during unloading, final indentation depth h_f can be solved by

$$F = Eh_f^2 \cdot f_u\left(\frac{Y}{E}, \frac{h_f}{h_m}, v, n, \theta\right) = 0. \tag{4.65}$$

Clearly, $\frac{h_f}{h_m}$ is a function of $\frac{Y}{E}$, v, n and θ.

Work done by the solid sample to the indenter during unloading W_u can be expressed:

$$W_u = \int_{h_f}^{h_m} F dh = Eh_m^3 \cdot \int_{h_f/h_m}^{1} x^2 f_u\left(\frac{Y}{E}, x, v, n, \theta\right) dx. \tag{4.66}$$

Integral $\int_{h_f/h_m}^{1} x^2 f_u(\frac{Y}{E}, x, v, n, \theta) dx$ in (4.66) is obviously a dimensionless function denoted as $f_{Wu}(\frac{Y}{E}, v, n, \theta)$. Therefore W_u can be:

$$W_u = \frac{Eh_m^3}{3} \cdot f_{Wu}\left(\frac{Y}{E}, v, n, \theta\right). \tag{4.67}$$

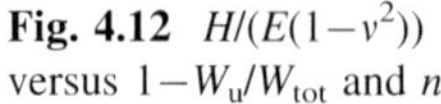

Fig. 4.12 $H/(E(1-v^2))$ versus $1-W_u/W_{tot}$ and n

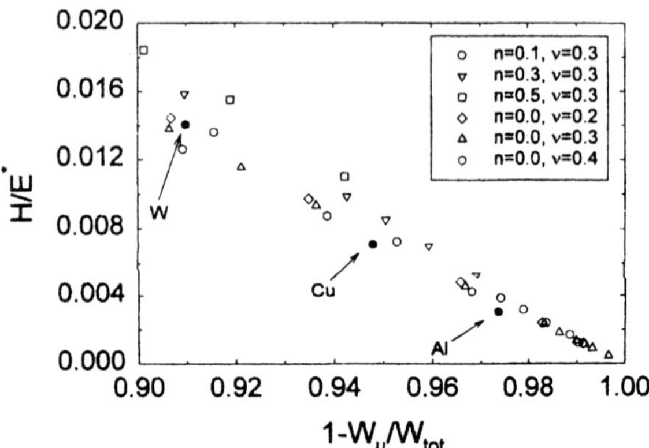

Consequently, work done on the indenter during unloading is proportional to h_m^3.

On the basis of loading work and unloading work, ratio of irreversible work $W_p(= W_{tot} - W_u)$ to total work W_{tot} for a complete loading–unloading cycle can be:

$$\frac{W_p}{W_{tot}} = \frac{W_{tot} - W_u}{W_{tot}} = 1 - \frac{f_{Wu}(Y/E,\ v,\ n,\ \theta)}{f_F(Y/E,\ v,\ n,\ \theta)}. \tag{4.68}$$

This ratio is independent of $\frac{h_m}{h_f}$.

Considering that $\frac{H}{E(1-v^2)}$ and $\frac{W_{tot}-W_u}{W_{tot}}$ are functions of $\frac{Y}{E}$, v, n and θ, finite element calculations can be used to evaluate $\frac{H}{E(1-v^2)}$ and $\frac{W_{tot}-W_u}{W_{tot}}$ for various material parameters but for the fixed θ $(= 68°)$. All calculation results are plotted in the form of $\frac{H}{E(1-v^2)}$ versus $1 - \frac{W_u}{W_{tot}}$ (Fig. 4.12, in which $E^* = E(1 - v^2)$). Fig. 4.12 shows that all data points lie nearly on a single curve, indicating an approximate one-to-one correspondence between $\frac{H}{E^*}$ and $1 - \frac{W_u}{W_{tot}}$. Consequently, this correspondence can be written as:

$$\frac{H}{E^*} \approx f_\theta\left(\frac{W_{tot} - W_u}{W_{tot}}\right), \tag{4.69}$$

where subscript θ = possible dependence on indenter angle, since such relationship was obtained for a particular indenter angle. Cheng replotted experimental data points for Cu (obtained by S. V. Hainsworth et al., see J. Mater. Res. 11, 1987 [1996]) and for W, Al (obtained by W. C. Oliver et al., see J. Mater. Res. 7, 1564 [1992]) shown in Fig. 4.12 in the form of $\frac{H}{E^*}$ versus $1 - \frac{W_u}{W_{tot}}$. There is a fair agreement between finite element calculations and experiments. It is particularly interesting that the behavior of conical indention exhibits characteristics of *self-similarity* in loading and unloading process.

The combination of dimensional analysis and *finite element calculations* provides insight into complicated phenomena.

4.4 Tensile Fracture of Solids

In 1950, an engine in the US ballistic missile Polaris exploded. After thorough investigation in situ and in the laboratory, specialists recognized that, for high strength materials, tensile strength criterion in which tensile stress or shear stress must be less than the corresponding strength, cannot be used as the sole criterion for judging the likelihood of damage. Another criterion number with dimension $FL^{-3/2}$ called *fracture toughness* characterizes critical behavior for fracture of materials.

Many scholars have tried to use classical elasto-plastic mechanics to analyze and illustrate fracture phenomena. However, those efforts have not been successful because material parameters in elasticity or plasticity are only *Young's modulus E*, Poisson's ratio v, and yield strength σ_s, while the dimension of E or σ_s is FL^{-2}, which differs from the dimension of fracture toughness $FL^{-3/2}$. Classical elasto-plastic mechanics must be improved in order to find a characteristic length important in deformation and fracture of materials that allows insight into the mechanism of fracture [4]. In the case of *extension of a crack* having one-dimensional strain, several classical assumptions provide a foundation for solving the problem:

1. Material is homogeneous and isotropic, without inner structures;
2. Material is elastic and elastic constants are Young's modulus E and Poisson's ratio v;
3. An original crack is a geometrical line segment, except for its length $2l$ and there is no other characteristic length in the problem (Fig. 4.13).

The problem is to calculate how great stress is required to make the crack begin to extend if tensile stress σ_0 increases gradually from 0. According to the assumptions, such tensile stress should be a function of related parameters that characterize the problem:

$$\sigma_0 = f(l,\ E,\ v). \tag{4.70}$$

Taking l and E as a unit system:

$$\frac{\sigma_0}{E} = f(v). \tag{4.71}$$

It shows that the tensile stress that makes the crack extend, does not relate to length of the original crack, but such conclusion is contrary to experimental facts.

Griffith (1921, 1924) introduced the concept "*surface energy*" to illustrate fracture in glass and his concept can be extended to illustrate fractures in other materials. Griffith considered that an increase in the area of a new surface accompanies the extension of a crack. The energy supplied to increase a unit surface area is defined as surface energy γ, having dimension F/L. Thus, stress σ_0 needed for extending the crack is:

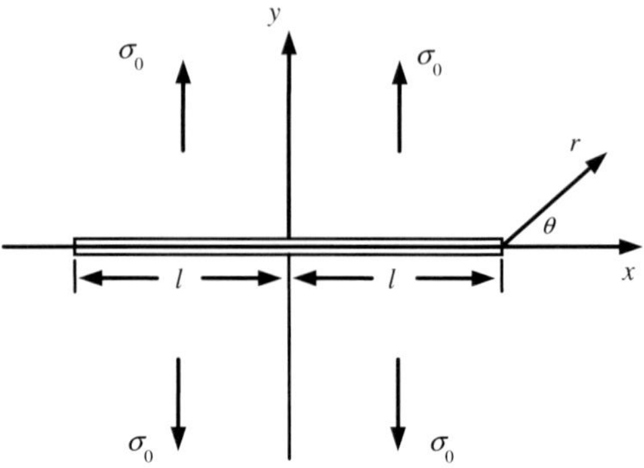

Fig. 4.13 Tensile fracture of solids

$$\sigma_0 = f(l,\ E,\ v,\ \gamma). \tag{4.72}$$

Similar to Griffith's treatment, it can be assumed that the extension of a crack is a conversion of elastic energy into surface energy. Thus, relationship (4.72) can be:

$$\frac{\sigma_0^2}{E} = f(l,\ v,\ \gamma). \tag{4.73}$$

Taking l and γ as a unit system and combining parameters related to elastic energy and surface energy produces dimensionless relationship:

$$\frac{\sigma_0^2/E}{\gamma/l} = f(v), \tag{4.74a}$$

or

$$\sigma_0^2 l = E\gamma \cdot f(v) \equiv C_G, \tag{4.74b}$$

where C_G = a constant determined by material properties E, v and γ. Relationship (4.74) shows that stress σ_0 necessary to make the crack grow is inversely proportional to $\sqrt{l}$, a result that is consistent with experiments.

In linearly elastic mechanics, stress distribution adjacent to the crack tip has the classical solution:

$$\begin{cases} \sigma_x = K_I \dfrac{1}{\sqrt{2\pi\pi}} \cdot \dfrac{1}{4}\left(3\cos\dfrac{\theta}{2} + \cos\dfrac{5\theta}{2}\right), \\[2ex] \sigma_y = K_I \dfrac{1}{\sqrt{2\pi\pi}} \cdot \dfrac{1}{4}\left(5\cos\dfrac{\theta}{2} - \cos\dfrac{5\theta}{2}\right), \\[2ex] \sigma_z = K_I \dfrac{1}{\sqrt{2\pi\pi}} \cdot \dfrac{1}{4}\left(-\sin\dfrac{\theta}{2} + \sin\dfrac{5\theta}{2}\right). \end{cases} \tag{4.75}$$

where $K_I = \sqrt{\pi l} \cdot \sigma_0$ is *stress intensity factor*.

Using the concept of stress intensity factor, the condition for crack extension, (4.74b) can be:

$$K_I = K_{IC}. \tag{4.76}$$

In other words, if stress intensity factor K_I reaches critical value K_{IC}, a crack extends further and K_{IC} = fracture toughness with dimension $FL^{-3/2}$. Combining (4.74b), (4.75), and (4.76) produces

$$K_{IC} = \sqrt{\pi E \gamma \cdot f(v)} = \sqrt{\pi C_G}. \tag{4.77}$$

Furthermore, linearly elastic mechanics proves that elastic energy related to crack extension necessary for length element dl is:

$$\frac{K_{IC}^2(1 - v^2)}{E} \cdot dl. \tag{4.78}$$

Analogous to Griffith's principle of *surface energy*, elastic energy is converted into energy required for the newly increased surface area $2\gamma \cdot dl$ and

$$\frac{K_{IC}^2(1 - v^2)}{E} \cdot dl = 2\gamma \cdot dl. \tag{4.79}$$

Writing K_{IC} as $\sqrt{\pi l} \cdot \sigma_{0c}$ according to (4.78), where σ_{0c} is regarded as tensile stress required to extend a crack with length l produces:

$$\sigma_{0c}^2 l = \frac{2}{\pi} \frac{\gamma E}{1 - v^2}, \tag{4.80a}$$

or

$$\frac{\pi(1 - v^2)\sigma_{0c}^2}{E} = \frac{2\gamma}{l}. \tag{4.80b}$$

To analyze the physical meaning of relationship (4.80b), $\frac{\sigma_{0c}^2}{E}$ on the left hand side characterizes elastic energy supplied to increase crack surface for an elastic body of unit volume and $\frac{2\gamma}{l}$ on the right hand side characterizes surface energy necessary for surface extension. Thus, the portion participating in energy conversion within the elastic body is just a layer of thickness δ beneath the crack surface.

To demonstrate the physical origin of the characteristic length δ involved in fracture toughness, fracture of *perfect crystal* can be selected as a typical case and there can be use of derived characteristic length δ and the non-local theory in linear elasticity of. Eringen [5], which extends classical continuum mechanics to deal with size effect of inner structure of materials. For a perfect crystal, the characteristic inner length δ is likely to be crystal constant a, the distance between two neighboring atoms. Eringen proposed that, for micro-heterogeneity, a macro stress–strain relationship can be derived by applying certain weighted averages to

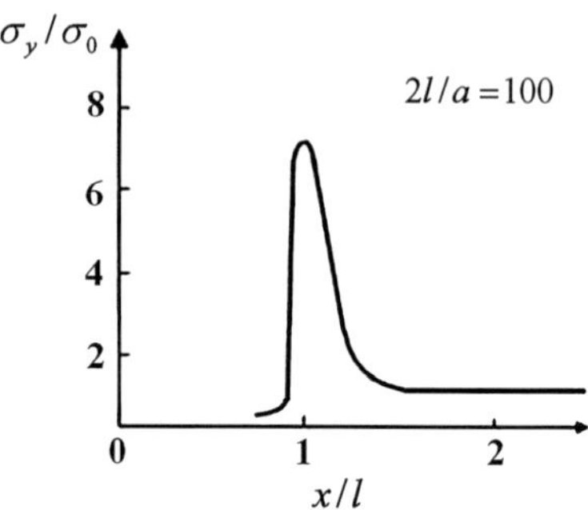

Fig. 4.14 Tensile stress distribution near crack tip

calculate distribution of σ_y along abscissa x near the crack tip (Fig. 4.14). The calculation shows that maximum of $\sigma_y(x, y)$ occurs at crack tip $x = l$ and $y = 0$, thus:

$$\frac{\sigma_y(l,\, 0)}{\sigma_0} = c \cdot \sqrt{\frac{2l}{a}}, \tag{4.81}$$

where $c = 0.73$. With sufficiently large x, the solution of $\sigma_y(x,\, y)$ for non-localized theory coincides with that of classical elasticity theory.

Using maximum tensile stress criterion for crack extension, $\sigma_y(l,\, 0) = \sigma_c$, where $\sigma_c = $ combination strength of perfect crystal, the condition for crack extension becomes

$$\sigma_0^2 l = \frac{\sigma_c^2 a}{2c^2} = C_G.$$

Here, constant C_G is completely determined by material properties σ_c and a, and derivation is based on the physical assumption of conversion from elastic energy to surface energy.

Summing up the above derivation, from the derivation based on Griffith's principle of surface energy:

$$C_G = \frac{2\gamma E}{\pi(1 - v^2)}, \tag{4.82}$$

and from the derivation based on Eringen's non-local theory of elasticity:

$$C_G = \frac{\sigma_c^2 a}{2c^2}. \tag{4.83}$$

The value of γ, surface energy provided by the physics of solids, can be substituted into (4.82) to obtain C_G and that C_G can be substituted into (4.83) to find σ_c:

Table 4.1 $\frac{\sigma_c}{E}$ for materials with different types of crystal lattices

Lattice type	Element	a (Å)	Non-local theory σ_c/E	Physics of solids σ_c/E
Face-centered cube	Al	2.86	0.18	–
Body-centered cube	Fe	2.48	0.18	0.23
Diamond	C	1.54	0.14	0.17
Close-packed cube	Zn	2.66	0.13	0.11

$$\sigma_c = \sqrt{\frac{4c^2\gamma E}{\pi(1 - v^2)a}}. \tag{4.84}$$

There is a fair agreement between this σ_c and the σ_c provided by the physics of solids. Relevant results regarding $\frac{\sigma_c}{E}$ for materials with different types of crystal lattices are in the following table: Table 4.1.

Regarding more complex metallic alloy materials, the characteristic inner length might be the distance between secondary particles in the case of dimple fracture and that inner size is probably the size of crystallite in the case of cleavage fracture with plastic deformation. Deeper understanding of the mechanism of fracture requires systematic experimental observations and theoretical analysis of typical cases under various conditions.

References

1. Miklowitz, J.: The Theory of Elastic Wave and Waveguides. Amsterdam, North-Holland Publ. Co., New York, London (1978)
2. Pochhammer, L.: Uber die Fortpflanzungsgeschwindigkeit kleiner Schwingungen in einem unbegrenzten isotropen Kreiscylinder. Journal für die reine und angewandte Mathematik. In zwanglosen Heften, Band 81, pp. 324-336, Berlin. (1876)
3. Cheng, Y.T., Cheng, C.M.: Analysis of indentation loading curves obtained using conical indenters. Phil. Mag. Lett. **77**(1), 39 47 (1998)
4. Cheng, C.M.: Mechanics of continuum and fracture. Adv. in Mech. **12**(2), 133–140 (1982). (in Chinese)
5. Eringen, A.C.: crystal lattice defects **7**, 109–130 (1977)

Chapter 5
Heat Conduction and Thermal Stress in Solids

Many problems involve effects that interconnect in particular ways and interfere with each other to form distinctive coupling. In other problems, coexisting effects can be considered to be decoupled. This chapter discusses problems related to heat conduction and thermal stress in solids. Generally, effects of heat conduction and thermal stress can be decoupled because the orders of magnitude of characteristic times for these effects differ considerably. The chapter discusses heat conduction, introduces a thermal stress–strain relationship and discusses the similarity law for thermal stress in solids.

5.1 Heat Conduction in Solids

In 1822, J. B. J. Fourier [1] presented a view that natural phenomena can be rigorously described only by equations in which all terms have the same dimensions. Fourier applied his view of dimensional homogeneity to a problem related to heat transfer in a bar with one end heated by a heat source and a side surface cooled by surrounding air. Fourier's discussion and analysis creates a fairly complete theory of heat transfer with two fundamental parameters. One parameter is coefficient of heat conduction, a physical property inherent in solid material that has a corresponding dimensionless parameter now known as the *Fourier number*. The other parameter, coefficient of heat transfer, induces a corresponding dimensionless parameter now known as the *Nusselt number* to reflect the nature of heat transfer between solid and surrounding air.

Fourier's experiments reveal laws *conductive and convective heat transfers*. His experimental data indicates that there are two kinds of rate of heat flux:

1. In a body

Rate of heat flux q in a body is the heat transferred through a body cross-section per unit area and per unit time, and q is proportional to normal temperature derivative at the cross-section:

Q.-M. Tan, *Dimensional Analysis*, DOI: 10.1007/978-3-642-19234-0_5,
© Springer-Verlag Berlin Heidelberg 2011

$$q \propto \frac{\partial T}{\partial n}.$$

For Fourier, the above equation cannot be a law of heat conduction because of its heterogeneous dimensions: dimension of q does not directly relate to dimension of $\frac{\partial T}{\partial n}$. Therefore, Fourier rewrites:

$$q = -\lambda \frac{\partial T}{\partial n}, \tag{5.1a}$$

where $\lambda =$ a proportional coefficient now called *heat conduction coefficient*, the dimension of which is ratio of dimension of q to dimension of $\frac{\partial T}{\partial n}$.

2. At interface between two bodies

Rate of heat flux q at an interface between two bodies with different temperatures is the heat exchanged at the interface per unit area and per unit time, and q is proportional to temperature difference between both bodies:

$$q \propto \Delta T.$$

Based on dimensional homogeneity requirement, Fourier rewrites:

$$q = \alpha \Delta T. \tag{5.1b}$$

Equation (5.1b) is Fourier's conception of a homogeneous law of convective heat transfer. α is *convective heat transfer coefficient*, the dimension of which is ratio of dimension of q to dimension of ΔT.

Based on heat transfer laws (5.1a) and (5.1b), Fourier created a mathematical theory of heat transfer that forms the basis for discussing heat conduction problems for prescribed boundary temperatures and for prescribed boundary temperature gradients.

5.1.1 Heat Conduction for Prescribed Boundary Temperatures

It is assumed that the characteristic lengths l, l', l'', …, can describe body geometry and that main material properties are density ρ, specific heat c and heat conductivity λ. It is also assumed that initial body temperature $=$ constant T_0 and boundary temperature at body surface $x_i = x_{iw}$ is an assigned function of space and time $T_w = T_{w0} \cdot f\left(\frac{x}{l}, \frac{y}{l}, \frac{z}{l}, \frac{t}{t_0}\right)$ for which characteristic temperature $= T_{w0}$, characteristic time $= t_0$ and $f =$ distribution function (Fig. 5.1). Actually, temperature difference $\theta \equiv T - T_0$ is a better variable than temperature T because temperature difference better represents problem essentials. Assigned initial temperature $T_0 =$ assigned initial temperature difference $\theta_0 = 0$ and assigned boundary temperature $T_w =$ assigned temperature difference θ_w:

Fig. 5.1 Heat conduction in solids

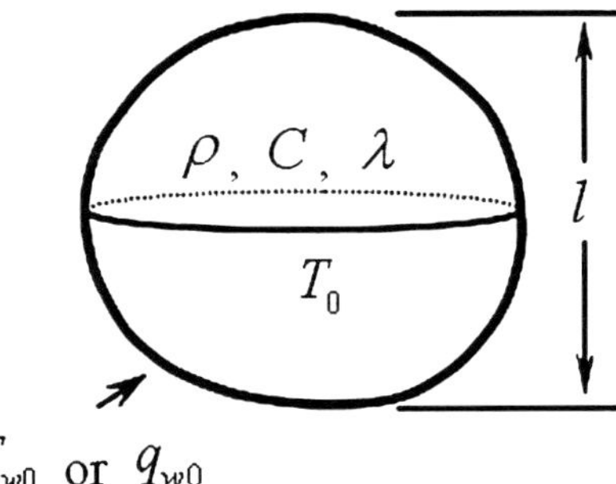

$$\theta_w = \theta_{w0} \cdot k\left(\frac{x_i}{l}, \frac{t}{t_0}\right), \tag{5.2}$$

where $\theta_{w0} = T_{w0} - T_0$ and $k =$ distribution function.

Using temperature difference, Fourier's heat conduction law is:

$$q = -\lambda \frac{\partial \theta}{\partial n}. \tag{5.3}$$

A problem related to heat conduction:

$$\begin{cases} \text{Equation: } \rho c \dfrac{\partial \theta}{\partial t} = \dfrac{\partial}{\partial x_i}\left(\lambda \dfrac{\partial \theta}{\partial x_i}\right), \\[2mm] \text{Initial condition: } t = 0 : \ \theta = 0, \\[2mm] \text{Boundary condition: } x_i = x_{iw} : \ \theta = \theta_{w0} \cdot k\left(\dfrac{x_i}{l}, \dfrac{t}{t_0}\right). \end{cases} \tag{5.4}$$

It is useful to discuss a simple case in which material is assumed to be homogeneous, i.e., density, specific heat, and heat conductivity are all constant. Thus, governing parameters in (5.4) are:

$$l, \ l', \ l'', \ \ldots, \ \frac{\rho c}{\lambda}, \ \theta_{w0}, \ t_0.$$

Temperature difference can be expressed as a function of spatial coordinates (x, y, and z) and time t, as well as governing parameters in (5.4):

$$\theta(x, y, z, t) = f\left(x, y, z, t; l, l', l'', \ldots; \frac{\rho c}{\lambda}; \theta_{w0}, t_0\right). \tag{5.5}$$

Taking l, θ_{w0}, and t_0 as a unit system, (5.5) reduces to dimensionless form:

$$\frac{\theta}{\theta_{w0}} = f\left(\frac{x}{l}, \frac{y}{l}, \frac{z}{l}, \frac{t}{t_0}; \frac{l'}{l}, \frac{l''}{l}, \ldots; \frac{\lambda t_0}{\rho c l^2}\right), \tag{5.6}$$

where the final dimensionless governing parameter is the Fourier number, denoted *Fo*:

$$Fo \equiv \frac{\lambda t_0}{\rho c l^2},\tag{5.7}$$

The dimensionless governing parameter (5.7) represents ratio of heat conductivity to heat inertia.

Analyzing these dimensionless independent variables in dimensionless relationship (5.6), if two conditions are satisfied:

1. Body geometries are similar: $\left(\frac{l'}{l}, \frac{l''}{l}, \ldots\right)_m = \left(\frac{l'}{l}, \frac{l''}{l}, \ldots\right)_p$,

2. Fourier numbers are identical: $(Fo)_m = (Fo)_p$,

then, at similar location and moment, $\left(\frac{x}{l}, \frac{y}{l}, \frac{z}{l}, \frac{t}{t_0}\right)_m = \left(\frac{x}{l}, \frac{y}{l}, \frac{z}{l}, \frac{t}{t_0}\right)_p$, dimensionless temperature differences are identical: $\left(\frac{\theta}{\theta_{w0}}\right)_m = \left(\frac{\theta}{\theta_{w0}}\right)_p$.

If the material in model and prototype is identical, second condition $(Fo)_m = (Fo)_p$ reduces:

$$\left(\frac{t_0}{l^2}\right)_m = \left(\frac{t_0}{l^2}\right)_p,$$

so,

$$\left(\frac{l_p}{l_m}\right)^2 = \left(\frac{t_{0p}}{t_{0m}}\right),$$

or

$$\alpha_{t_0} = \alpha_l^2.$$

Because similar time is required:

$$\left(\frac{t}{t_0}\right)_m = \left(\frac{t}{t_0}\right)_p, \quad \text{or} \quad \frac{t_p}{t_m} = \frac{t_{0p}}{t_{0m}},$$

a requirement follows:

$$\alpha_t = \alpha_l^2.$$

Based on requirements of $\left(\frac{l'}{l}, \frac{l''}{l}, \ldots\right)_m = \left(\frac{l'}{l}, \frac{l''}{l}, \ldots\right)_p$ and $(Fo)_m = (Fo)_p$, small scale model testing or theoretical calculations can be used to find dimensionless distribution of temperature difference $\frac{\theta}{\theta_{w0}}$, which gives temperature difference at similar spatial and temporal coordinates in the prototype.

If dimensionless density, specific heat and heat conductivity are:

$$\begin{cases} \dfrac{\rho}{\rho_0} = m\left(\dfrac{x}{l}, \dfrac{y}{l}, \dfrac{z}{l}\right), \\[2ex] \dfrac{c}{c_0} = n\left(\dfrac{x}{l}, \dfrac{y}{l}, \dfrac{z}{l}; \dfrac{\theta}{\theta_{w0}}\right), \\[2ex] \dfrac{\lambda}{\lambda_0} = q\left(\dfrac{x}{l}, \dfrac{y}{l}, \dfrac{z}{l}; \dfrac{\theta}{\theta_{w0}}\right), \end{cases}$$

where m, n and $q = $ given corresponding distribution functions, discussion can be analogous to discussion in a homogeneous case and similar results hold for heterogeneous materials. However, distribution functions m, n and q must be identical for model and prototype.

5.1.2 Heat Conduction for Prescribed Temperature Gradient at Boundary

It is assumed that body geometry represented by characteristic lengths l, l', l'',... is given and that material properties represented by density ρ, specific heat c and heat conductivity λ are given. It is also assumed that initial body temperature $= T_0$ and heat transfer condition at the body surface $x_i = x_{iw}$ is:

$$\lambda\frac{\partial T}{\partial n} = \alpha(T_w - T_e), \tag{5.8}$$

where $T_w = $ body surface temperature, $T_e = $ environmental temperature, $\alpha = $ heat transfer coefficient, $\frac{\partial}{\partial n} = $ normal derivative at body surface and T_e and α are given. T_w is unknown and regarded as being part of the temperature distribution that is sought. Heat transfer coefficient α depends on material properties and motion behavior of the surrounding medium, and should be given beforehand.

Using temperature difference θ as a dependent variable allows a heat conduction problem to be formulated:

$$\begin{cases} \text{Equation: } \rho c\dfrac{\partial\theta}{\partial t} = \dfrac{\partial}{\partial x_i}\left(\lambda\dfrac{\partial\theta}{\partial x_i}\right); \\[2ex] \text{Initial condition: } t = 0 : \theta = 0; \\[2ex] \text{Boundary condition: } x_i = x_{iw} : \lambda\dfrac{\partial\theta}{\partial n} = \alpha(\theta_w - \theta_e), \end{cases} \tag{5.9}$$

where subscript $w = $ boundary surface and environmental temperature difference θ_e and heat transfer coefficient α are prescribed functions:

$$
\begin{cases}
\theta_e = \theta_{e0} \cdot h\left(\dfrac{t}{t_0}\right), \\[2ex]
\alpha = \alpha_0 \cdot g\left(\dfrac{x}{l}, \dfrac{y}{l}, \dfrac{z}{l}, \dfrac{\theta}{\theta_{e0}}\right),
\end{cases}
\tag{5.10}
$$

where θ_{e0} = characteristic temperature difference, h = distribution function of θ_e, α_0 = characteristic heat transfer coefficient and g = distribution function of α.

For homogeneous materials, all properties are constant and body temperature difference can be:

$$
\theta = f\left(x, \ y, \ z, \ t; \ l, \ l', \ \ldots; \ \frac{\rho c}{\lambda}, \ \frac{\alpha_0}{\lambda}, \ \theta_{e0}, \ t_0\right).
\tag{5.11}
$$

Taking l, θ_{e0} and t_0 as a unit system (5.11) reduces to dimensionless expression:

$$
\frac{\theta}{\theta_{e0}} = f\left(\frac{x}{l}, \ \frac{y}{l}, \ \frac{z}{l}, \ \frac{t}{t_0}; \ \frac{l'}{l}, \ \frac{l''}{l}, \ \ldots; \ \frac{\lambda t_0}{\rho c l^2}; \ \frac{\alpha_0 l}{\lambda}\right).
\tag{5.12}
$$

On the right hand side of (5.12), there are two dimensionless parameters in the function f, *Fourier number* $Fo \equiv \frac{\lambda t_0}{\rho c l^2}$ and *Nusselt number* $Nu \equiv \frac{\alpha_0 l}{\lambda}$, where *Nusselt number* represents a ratio of two thermal characteristics, i.e., heat transfer at the body surface and heat conduction in the body.

If conditions in the modeling tests are satisfied:

1. Body geometries are similar: $\left(\frac{l'}{l}, \frac{l''}{l}, \ldots\right)_m = \left(\frac{l'}{l}, \frac{l''}{l}, \ldots\right)_p$;
2. Fourier numbers are the same: $(Fo)_m = (Fo)_p$;
3. Nusselt numbers are the same: $(Nu)_m = (Nu)_p$;
4. Distribution functions g and h in the boundary condition are the same, then at similar points in space and time

$$
\left(\frac{x}{l}, \ \frac{y}{l}, \ \frac{z}{l}, \ \frac{t}{t_0}\right)_m = \left(\frac{x}{l}, \ \frac{y}{l}, \ \frac{z}{l}, \ \frac{t}{t_0}\right)_p,
$$

dimensionless temperature differences are the same:

$$
\left(\frac{\theta}{\theta_{e0}}\right)_m = \left(\frac{\theta}{\theta_{e0}}\right)_p.
\tag{5.13}
$$

Discussion of heterogeneous materials and discussion of homogeneous materials are similar.

5.2 Thermal Stress in Elastic Bodies

In solids, thermal expansion causes *thermal stress* due to heterogeneous temperature distribution. Temperature is the manifestation of random thermal motion of molecules, atoms and ions. Generally, random thermal motion influence macro-displacements and macro-displacements produce stress and strain. However, even if there is thermal expansion, there is no thermal stress unless there is resistance to macro-displacement or deformation in body elements. If macro-displacements are constrained, thermal expansion produces thermal stress.

Local temperature elevation in an elastic body causes local thermal expansion and local variation in stress and strain that propagates outwards and affects neighboring regions in the form of elastic waves. Elastic wave propagation and heat conduction possess different *characteristic time* scales.

The order of magnitude of propagation velocity in elastic waves can be estimated by $(E/\rho)^{1/2}$, where $E = $ *Young's modulus* and $\rho = $ density. The time needed to recover from stress state caused by elastic wave disturbance to equilibrium state has roughly the same order of magnitude as that of time $t_{e.w.}$ needed for propagation of elastic waves from local disturbance origin to body boundary. If characteristic body length $= l$ and order of magnitude of time $t_{e.w.}$ is about:

$$t_{e.w.} \approx \frac{l}{(E/\rho)^{1/2}}. \tag{5.14}$$

If characteristic time $t_{e.w.}$ that represents propagation of elastic waves, has the same order of magnitude as that of the characteristic time that represents heat conduction $t_{h.c.}$, effects of heat conduction and thermal stress are coupled and should be considered together. To estimate *characteristic time for heat conduction* $t_{h.c.}$, if characteristic body length $= l$, Fourier's heat conduction equation is:

$$\rho c \frac{\partial \theta}{\partial t} = \lambda \frac{\partial}{\partial x_l} \left(\frac{\partial \theta}{\partial x_l} \right),$$

characteristic time for heat conduction $t_{h.c.}$ should be:

$$t_{h.c.} = f\left(\frac{\rho c}{\lambda}, l \right).$$

Taking $\frac{\rho c}{\lambda}$ and l as a unit system generates:

$$t_{h.c.} \approx \frac{\rho c l^2}{\lambda}. \tag{5.15}$$

In general, it is easy to see that $t_{e.w.} \ll t_{h.c.}$, i.e.,

$$\frac{l}{(E/\rho)^{1/2}} < < \frac{\rho c l^2}{\lambda}. \tag{5.16}$$

For example, if a body made of steel has characteristic length $l = 1\ m$ with density $\rho = 7.8 \times 10^3 kg/m^3$, specific heat $c = 0.50 \times 10^3 N \cdot m/(kg \cdot K)$ and heat conductivity $\lambda = 50N/(\sec \cdot K)$, characteristic time for heat conduction is about:

$$t_{h.c.} \approx \frac{\rho c l^2}{\lambda} = \frac{7.8 \times 10^3 \times 0.50 \times 10^3 \times 1^2}{50}\sec = 0.78 \times 10^5 \sec.$$

Based on *Young's modulus* of steel $E = 2.1 \times 10^{11} kg/(\sec^2 \cdot m)$, propagation time for elastic waves is about:

$$t_{e.w.} \approx \frac{l}{(E/\rho)^{1/2}} = \frac{1}{(2.1 \times 10^{11}/7.8 \times 10^3)^{1/2}}\sec$$
$$= \frac{1m}{0.52 \times 10^4 m/\sec} = 1.9 \times 10^{-4}\sec.$$

Ratio of two characteristic times:

$$\frac{t_{e.w.}}{t_{h.c.}} = \frac{1.9 \times 10^{-4}\sec}{0.78 \times 10^5 \sec} \cong 2 \times 10^{-9}.$$

In this case, two characteristic times differ by nearly 9 orders of magnitude. Variation of stress and strain caused by local temperature increasingly propagates quickly throughout the body. However, temperature variation due to heat conduction is much slower than variation of stress and strain. Heat conduction is substantially slower than mechanical disturbance propagation, so these characteristics can be viewed as being decoupled.

Generally, a problem of thermal stress can be solved in two steps.

Step 1. Temperature distribution throughout a body is understood by solving the problem of heat conduction.

Step 2. Elastic stress distribution caused by thermal expansion is understood by focusing on an elastic body with given temperature distribution.

However, there are exceptions. For example, if heating rate is so high that times $t_{e.w.}$ and $t_{h.c.}$ have the same order of magnitude, heat conduction is regarded as being a form of wave propagation. Such a case appears at the end of this chapter.

5.2.1 Thermal Elastic Constitutive Relationship

Conducting dimensional analysis for a thermal stress problem requires knowing elastic constitutive relationship applicable to thermal expansion effect. In general cases, Hooke's law covers elastic constitutive relationship:

$$\text{Normal direction} \qquad\qquad \text{Tangential direction}$$

$$
\begin{aligned}
\varepsilon_{11} &= \frac{\sigma_{11} - v(\sigma_{22} + \sigma_{33})}{E}, & \varepsilon_{12} &= \frac{2(1+v)}{E}\sigma_{12}, \\[2mm]
\varepsilon_{22} &= \frac{\sigma_{22} - v(\sigma_{33} + \sigma_{11})}{E}, & \varepsilon_{23} &= \frac{2(1+v)}{E}\sigma_{23}, \\[2mm]
\varepsilon_{33} &= \frac{\sigma_{33} - v(\sigma_{11} + \sigma_{22})}{E}, & \varepsilon_{31} &= \frac{2(1+v)}{E}\sigma_{31},
\end{aligned}
\tag{5.17}
$$

where strain ε_{ij} is caused solely by stress σ_{ij} and is unaffected by thermal expansion.

Assuming total strain ε_{ij} to be a linear sum of two parts, one part caused by stress $\varepsilon_{ij}^{\sigma}$ and the other part caused by temperature difference $\varepsilon_{ij}^{\theta}$:

$$\varepsilon_{ij} = \varepsilon_{ij}^{\sigma} + \varepsilon_{ij}^{\theta}. \tag{5.18}$$

Because stress induced by temperature difference $\varepsilon_{ij}^{\theta}$ is merely a kind of volumetric strain, temperature difference affects normal strains but does not affect shear strains. Normal strains are assumed to be proportional to temperature difference, so:

$$\text{Normal direction}: \ \varepsilon_{11}^{\theta} = \varepsilon_{22}^{\theta} = \varepsilon_{33}^{\theta} = \alpha_e\theta, \tag{5.19a}$$

$$\text{Tangential direction}: \ \varepsilon_{12}^{\theta} = \varepsilon_{23}^{\theta} = \varepsilon_{31}^{\theta} = 0, \tag{5.19b}$$

where $\alpha_e = $ *linear expansion coefficient* equal to 1/3 of *volumetric expansion coefficient* β.

Elastic constitutive relationship applicable to thermal expansion effect, i.e., *thermal-elasticity constitutive relationship* can be:

$$\text{Normal direction} \qquad\qquad \text{Tangential direction}$$

$$
\begin{aligned}
\varepsilon_{11} &= \frac{\sigma_{11} - v(\sigma_{22} + \sigma_{33})}{E} + \frac{\beta\theta}{3}, & \varepsilon_{12} &= \frac{2(1+v)}{E}\sigma_{12}, \\[2mm]
\varepsilon_{22} &= \frac{\sigma_{22} - v(\sigma_{33} + \sigma_{11})}{E} + \frac{\beta\theta}{3}, & \varepsilon_{23} &= \frac{2(1+v)}{E}\sigma_{23}, \\[2mm]
\varepsilon_{33} &= \frac{\sigma_{33} - v(\sigma_{11} + \sigma_{22})}{E} + \frac{\beta\theta}{3}, & \varepsilon_{31} &= \frac{2(1+v)}{E}\sigma_{31}.
\end{aligned}
\tag{5.20}
$$

Using strain ε_{ij} and temperature difference θ to express stress σ_{ij}:

Normal direction:

$$
\begin{aligned}
\sigma_{11} &= 2G\cdot\left[\varepsilon_{11} + \frac{v}{1-2v}(\varepsilon_{11} + \varepsilon_{22} + \varepsilon_{33}) - \frac{1+v}{1-2v}\cdot\frac{\beta\theta}{3}\right], \\[2mm]
\sigma_{22} &= 2G\cdot\left[\varepsilon_{22} + \frac{v}{1-2v}(\varepsilon_{11} + \varepsilon_{22} + \varepsilon_{33}) - \frac{1+v}{1-2v}\cdot\frac{\beta\theta}{3}\right], \\[2mm]
\sigma_{33} &= 2G\cdot\left[\varepsilon_{33} + \frac{v}{1-2v}(\varepsilon_{11} + \varepsilon_{22} + \varepsilon_{33}) - \frac{1+v}{1-2v}\cdot\frac{\beta\theta}{3}\right];
\end{aligned}
\tag{5.21a}
$$

Tangential direction:

$$\sigma_{12} = G\varepsilon_{12},$$
$$\sigma_{23} = G\varepsilon_{23}, \qquad (5.21b)$$
$$\sigma_{31} = G\varepsilon_{31}.$$

In (5.21a) and (5.21b), G is shear modulus: $G = \frac{E}{2(1+v)}$.

5.2.2 Thermal Stress in Solids

It is possible to deal with a thermal stress problem having an assigned boundary temperature:

$$x_i = x_{iw} : \ \theta = \theta_{w0} \cdot k\left(\frac{x_i}{l}, \frac{t}{t_0}\right), \qquad (5.22)$$

where θ_{w0} = characteristic temperature difference at the boundary, k = distribution function, l = characteristic body length and t_0 = characteristic time of temperature distribution function k. Therefore, parameters presented in boundary conditions are l, θ_{w0}, and t_0.

Similar analysis can be applied to assigned heat transfer at the boundary.

As stated previously, a problem of thermal stress can be decoupled and solved in two steps:

Step 1. Dimensionless temperature difference $\frac{\theta}{\theta_{w0}}$ is found out first.

Step 2. Thermal stress distribution σ_{ij} is discussed based on the given dimensionless temperature difference $\frac{\theta}{\theta_{w0}}$.

In Step 2, using the given $\frac{\theta}{\theta_{w0}}$, thermal stress–strain relationship can be dimensionless:

Normal direction:

$$\frac{\sigma_{11}}{G} = 2 \cdot \left[\varepsilon_{11} + \frac{v}{1-2v}(\varepsilon_{11} + \varepsilon_{22} + \varepsilon_{33}) - \frac{1+v}{1-2v} \cdot \frac{\beta\theta_{w0}}{3} \cdot \frac{\theta}{\theta_{w0}}\right],$$
$$\frac{\sigma_{22}}{G} = 2 \cdot \left[\varepsilon_{22} + \frac{v}{1-2v}(\varepsilon_{11} + \varepsilon_{22} + \varepsilon_{33}) - \frac{1+v}{1-2v} \cdot \frac{\beta\theta_{w0}}{3} \cdot \frac{\theta}{\theta_{w0}}\right], \qquad (5.23a)$$
$$\frac{\sigma_{33}}{G} = 2 \cdot \left[\varepsilon_{33} + \frac{v}{1-2v}(\varepsilon_{11} + \varepsilon_{22} + \varepsilon_{33}) - \frac{1+v}{1-2v} \cdot \frac{\beta\theta_{w0}}{3} \cdot \frac{\theta}{\theta_{w0}}\right];$$

Tangential direction:

$$\frac{\sigma_{12}}{G} = \varepsilon_{12},$$
$$\frac{\sigma_{23}}{G} = \varepsilon_{23}, \qquad (5.23b)$$
$$\frac{\sigma_{31}}{G} = \varepsilon_{31}.$$

To summarize, three types of parameters govern thermal stress problem.

Type 1. Body geometry: characteristic lengths l, l', $\cdots$;

Type 2. Material properties: elastic constants G, v, and thermal volumetric expansion coefficient β;

Type 3. Boundary conditions: characteristic temperature difference θ_{w0} and characteristic time t_0.

Dimensionless stress–strain relationship (5.23a), (5.23b) make it possible to find β and θ_{w0} in the form of product $\beta\theta_{w0}$. Therefore, stress distribution can be:

$$\sigma_{ij} = f_{ij}(x,\ y,\ z,\ t;\ l,\ l',\ \ldots;\ G,\ v;\ \beta\theta_{w0};\ t_0). \tag{5.24}$$

Taking l, G and t_0 as a unit system produces dimensionless relationship:

$$\frac{\sigma_{ij}}{G} = f_{ij}\left(\frac{x}{l},\ \frac{y}{l},\ \frac{z}{l},\ \frac{t}{t_0};\ \frac{l'}{l},\ \ldots;\ v;\ \beta\theta_{w0}\right). \tag{5.25}$$

The problem possesses property of linearity and stress is proportional to characteristic temperature difference θ_{w0}. Thus:

$$\frac{\sigma_{ij}}{G} = \beta\theta_{w0} \cdot f_{ij}\left(\frac{x}{l},\ \frac{y}{l},\ \frac{z}{l},\ \frac{t}{t_0};\ \frac{l'}{l},\ \ldots;\ v\right). \tag{5.26}$$

Maximum thermal stress in elastic body:

$$\frac{\sigma_{ij}^{m}}{G} = G\beta\theta_{w0} \cdot f_{ij}\left(\frac{l'}{l},\ \ldots;\ v\right). \tag{5.27}$$

If model and prototype are geometrically similar and have the same Poisson's ratio and the same distribution functions of temperature difference at the boundary, the same dimensionless stress exists at similar location and time:

$$at\ \left(\frac{x}{l},\ \frac{y}{l},\ \frac{z}{l},\ \frac{t}{t_0}\right)_m = \left(\frac{x}{l},\ \frac{y}{l},\ \frac{z}{l},\ \frac{t}{t_0}\right)_p:$$
$$\left(\frac{\sigma_{ij}}{G\beta\theta_{w0}}\right)_m = \left(\frac{\sigma_{ij}}{G\beta\theta_{w0}}\right)_p. \tag{5.28}$$

A body's thermal response time and mechanical response time may be close in extremely intense heating with extremely short characteristic time, such as in intense chemical reaction, thermonuclear fusion or laser-radiation. In such a case, Fourier's conduction law is ineffective because thermal disturbance propagation depends on interactions between electron and electron, electron and phonon, phonon and phonon, etc.

Thermal disturbances that propagate with finite velocity are *thermal waves*. Relaxation time needed for the medium to recover from disturbed state to local equilibrium state is about $10^{-9} \sim 10^{-14}$ sec. Correspondingly, thermal wave propagation velocity is about $10^2 \sim 10^5$ m/sec, which approximates the order of magnitude of propagation velocity of elastic waves. In such a case, heat conduction effect and thermal stress effect are coupled. Inertia effect caused by heat conduction needs to be considered because decoupling is no longer valid [2].

References

1. Fourier. J.B.J.: Analytic theory of heat, Dover publications, Inc. Newyork (1955)
2. Duan, Z.P., Fu, Y.S.: On the theory of thermal waves. Adv. in Mech. **22**(4), 433–448 (1992). (in Chinese)

Chapter 6
Problems of Coupling Fluid Motion and Solid Deformation

Some natural phenomena or engineering problems relate to fluid–solid interaction that involves *coupling of fluid motion and solid deformation*. Typical examples are water hammers and fluttering airfoils; more complex examples include deformation of blood vessels, blood flow, and pulse phenomena in the human body. This chapter uses dimensional analysis to discuss cases related to a water hammer, elastic bearings, fluttering airfoils and *vortex-excited vibration* of heat exchangers.

6.1 Water Hammers

Abrupt closure of a valve or breakdown of a pipeline system source causes rapid change in flow velocity of water and corresponding pressure surge. Such action causes a *water hammer* effect wherein a pressure wave propagates along the pipe and mechanical impact produces sound. Pressure variation causes simultaneous variation in water volume and elastic deformation of pipe, resulting in pressure wave propagation and determining pressure surge amplitude. Such amplitude indicates *water hammer intensity* and is an important parameter in engineering design.

6.1.1 Wave Velocity of Pressure Waves

A *water hammer* pressure wave is a compressible flow phenomenon in elastic water. The pipe expands due to water pressure surge so that the cross-section of pipe A varies with water pressure:

$$A = A(p). \tag{6.1}$$

Density of water ρ also varies with pressure according to the state of water:

$$\rho = \rho(p). \tag{6.2}$$

Q.-M. Tan, *Dimensional Analysis*, DOI: 10.1007/978-3-642-19234-0_6,

The water hammer problem can be formulated as a one-dimensional compressible flow in an elastic pipe of varying cross-section. Such flow satisfies conservation laws for mass and momentum:

$$
\begin{cases}
\text{Conversation of mass: } \dfrac{\partial \rho A}{\partial t} = -\dfrac{\partial(\rho A v)}{\partial x}, \\[3mm]
\text{Conversation of momentum: } \rho\left(\dfrac{\partial v}{\partial t} + v\dfrac{\partial v}{\partial x}\right) = -\dfrac{\partial p}{\partial x},
\end{cases}
\tag{6.3}
$$

where pressure p, density ρ, velocity v and cross-section of pipe A are functions of space and time coordinates x and t.

Because pipe deformation is elastic and water compressibility is low, variation of cross-section ΔA and variation of density of water $\Delta\rho$ are small compared with undisturbed basic state:

$$
\begin{cases}
p = p_0, \ \rho = \rho(p_0) = \rho_0, \ v = v_0; \\
A = A(p_0) = A_0
\end{cases}
$$

Thus:

$$
\frac{\Delta A}{A_0} \ll 1, \quad \frac{\Delta\rho}{\rho_0} \ll 1.
$$

Abrupt velocity change $\Delta v(= v_0) \ll$ speed of sound in water c_w:

$$
\frac{\Delta v}{c_w} \ll 1.
$$

Equations in (6.3) can be linearized:

$$
\begin{cases}
\text{Mass conservation : } \left[\dfrac{d(\rho A)}{dp}\right]_{p=p_0} \cdot \dfrac{\partial p}{\partial t} = -\rho_0 A_0 \dfrac{\partial v}{\partial x}, \\[3mm]
\text{Momentum conservation : } \rho_0 \dfrac{\partial v}{\partial t} = -\dfrac{\partial p}{\partial x}.
\end{cases}
\tag{6.4}
$$

To eliminate v in (6.4), differentiating equation of mass conservation with respect to t produces:

$$
\left[\frac{d(\rho A)}{dp}\right]_{p=p_0} \cdot \frac{\partial^2 p}{\partial t^2} = -\rho_0 A_0 \frac{\partial}{\partial x}\left(\frac{\partial v}{\partial t}\right).
$$

By using equation of momentum conservation, factor $-\rho_0\frac{\partial v}{\partial t}$ on the right hand side of the above equation can be replaced by $\frac{\partial p}{\partial x}$:

$$
\frac{\partial^2 p}{\partial t^2} = \frac{A_0}{[d(\rho A)/dp]_{p=p_0}} \cdot \frac{\partial^2 p}{\partial x^2}.
$$

Since $A(p)$ and $\rho(p)$ are functions that increase with p, the coefficient of the right hand side of the above equation is positive:

$$\frac{A_0}{[d(\rho A)/dp]_{p=p_0}} > 0,$$

so it can be regarded as:

$$c^2 = \frac{A_0}{[d(\rho A)/dp]_{p=p_0}}. \tag{6.5}$$

Thus, pressure p satisfies wave equation:

$$\frac{\partial^2 p}{\partial t^2} = c^2 \cdot \frac{\partial^2 p}{\partial x^2}, \tag{6.6}$$

where c = pressure wave propagation velocity in the water hammer, which can be called *sound velocity*.

Examining the reciprocal of c^2 allows better understanding of roles played by water pipe elasticity and water *compressibility*:

$$\frac{1}{c^2} = \left[\frac{d(\rho A)}{dp}\right]_{p=p_0} \bigg/ A_0.$$

Divided by density of undisturbed water ρ_0 produces:

$$\frac{1}{\rho_0 c^2} = \left[\rho \frac{dA}{dp} + A \frac{d\rho}{dp}\right]_{p=p_0} \bigg/ (\rho_0 A_0),$$

or

$$\frac{1}{\rho_0 c^2} = \left[\frac{1}{A}\frac{dA}{dp} + \frac{1}{\rho}\frac{d\rho}{dp}\right]_{p-p_0}. \tag{6.7}$$

The right hand side of relationship (6.7) is the sum of $\left[\frac{1}{A}\frac{dA}{dp}\right]_{p=p_0}$ that characterizes pipe elasticity and $\left[\frac{1}{\rho}\frac{d\rho}{dp}\right]_{p=p_0}$ that characterizes water compressibility. $\left[\frac{1}{\rho}\frac{d\rho}{dp}\right]_{p=p_0}$ is the reciprocal of the compression modulus of water. With rigid pipe, A is irrelevant to p and sound velocity c reduces to *sound velocity* in water c_w:

$$c_w = \left(\frac{dp}{d\rho}\right)_{p=p_0}^{\frac{1}{2}} = \left(\frac{K}{\rho_0}\right)^{\frac{1}{2}}, \tag{6.8}$$

where K = compression modulus of water $\left[\rho\left(\frac{dp}{d\rho}\right)\right]_{p=p_0}$.

6.1.2 Intensity of Water Hammer

Water hammer intensity is defined by amplitude of pressure wave Δp, which depends on abrupt velocity change Δv, inertia and elasticity of pipe and, inertia and compressibility of water. Parameters that govern pipe deformation are diameter D, wall thickness δ, density ρ, *Young's modulus* E, and Poisson's ratio v. Parameters that govern water deformation are density ρ_0 and compression modulus K. Thus, intensity of water hammer Δp:

$$\Delta p = f(\Delta v;\ D,\ \delta,\ \rho,\ E,\ v;\ \rho_0,\ K). \tag{6.9}$$

Taking D, ρ_0 and K as a unit system, reduces (6.9) to dimensionless relationship:

$$\frac{\Delta p}{K} = f\left(\frac{\Delta v}{\sqrt{K/\rho_0}};\ \frac{\delta}{D},\ \frac{E}{K},\ v;\ \frac{\rho}{\rho_0}\right). \tag{6.10}$$

From the viewpoint of physics, a problem with small disturbances possess property of linearity and intensity of water hammer Δp is proportional to abrupt velocity change Δv. Thus:

$$\frac{\Delta p}{K} = \frac{\Delta v}{\sqrt{K/\rho_0}} \cdot f\left(\frac{\delta}{D},\ \frac{E}{K},\ v;\ \frac{\rho}{\rho_0}\right). \tag{6.11a}$$

Because $\dfrac{K}{\sqrt{K/\rho_0}} = \rho_0\sqrt{K/\rho_0} = \rho_0 c_w$, expression (6.11a) can be:

$$\Delta p = \rho_0 c_w \Delta v \cdot f\left(\frac{\delta}{D},\ \frac{E}{K},\ v;\ \frac{\rho}{\rho_0}\right). \tag{6.11b}$$

If pipe deformation is ignored, expression (6.11b) reduces to a basic relationship for a sound wave in water:

$$\Delta p = \rho_0 c_w \Delta v, \tag{6.12}$$

where $c_w = $ *sound velocity* in water, and $\rho_0 c_w = $ *acoustic impedance* of water.

To consider pipe deformation, two cases can be discussed according to relative wall thickness $\frac{\delta}{D}$.

Case 1. Pipe with thin wall: A water pipe with a relatively thin wall can be seen as an axisymmetrical rotary shell having deformation composed of membrane tension and shell flexure. Material property that resists tension is tension rigidity $E\delta$ and material property that resists flexure is flexural rigidity $\frac{E\delta^3}{(1-v^2)KD^3}$. Pipe inertia $= \rho\delta$ and water inertia $= \rho_0 D$. Recombining dimensionless independent variables $\frac{\delta}{D}, \frac{E}{K}, v$ and $\frac{\rho}{\rho_0}$ forms a set of new variables $\frac{E\delta}{KD}, \frac{E\delta^3}{(1-v^2)KD^3}, v$ and $\frac{\rho\delta}{\rho_0 D}$. Thus, intensity of water hammer Δp can be:

$$\Delta p = \rho_0 c_w \Delta v \cdot f\left(\frac{E\delta}{KD}, \ \frac{E\delta^3}{(1-v^2)KD^3}, \ v; \ \frac{\rho\delta}{\rho_0 D}\right). \tag{6.13}$$

This case allows two possibilities:

Possibility 1. If $\frac{\delta}{D}$ is small enough, membrane tension dominates pipe deformation and shell flexure can be ignored:

$$\Delta p = \rho_0 c_w \Delta v \cdot f\left(\frac{E\delta}{KD}, \ v, \ \frac{\rho\delta}{\rho_0 D}\right). \tag{6.14}$$

Possibility 2. If $\frac{\delta}{D}$ is large enough, shell flexure dominates pipe deformation and membrane tension can be ignored:

$$\Delta p = \rho_0 c_w \Delta v \cdot f\left(\frac{E\delta^3}{(1-v^2)KD^3}, \ v; \ \frac{\rho\delta}{\rho_0 D}\right). \tag{6.15}$$

Case 2. Pipe with thick wall: In a channel in rock, used for transporting water, a pressure wave reaches the interface between water and rock and refracts into rock. When refractive wave reaches rock surface, the reflection effect from the rock surface is weakened by thick rock, so $\frac{\delta}{D}$ is not a governing parameter. Based on elasticity theory, intensity of the pressure wave that refracts from water to rock, depends on ratio of acoustic impedance of rock to acoustic impedance of water $\sqrt{\frac{\rho E}{\rho_0 K}}$ and depends on Poisson's ratio v.

When a valve closes abruptly, maximum pressure $(\Delta p)_{\max}$ is exerted on the valve:

$$\Delta p_{\max} = \rho_0 c_w \Delta v \cdot f\left(\frac{\rho E}{\rho_0 K}, \ v\right), \tag{6.16a}$$

and maximum stress $\sigma_{\max}$ is exerted on the rock:

$$\sigma_{\max} = \rho_0 c_w \Delta v \cdot f\left(\frac{\rho E}{\rho_0 K}, \ v\right), \tag{6.16b}$$

where particular forms of functions f and g can be determined by model testing or theoretical analysis.

6.2 Elastic Bearings with Liquid Lubricant

In contrast with viscous effect of lubricant on bearings designed for light loading (Sect. 3.1), axle and bearing surfaces may have serious wear if loading is heavy enough to reduce minimum space between axle and bearing. In such a case, elastic deformation of axles and bearings needs consideration. It is assumed that bearing radius $= R$, axle radius $= r$, lubricant viscosity $= \mu$, loading exerted on axle $= W$, relative velocity between two sliding surfaces $= v$ and mean space between two sliding surfaces $= R - r$. In general, relative velocity is too low to make ratio of inertia effect to *viscosity effect*, $\mathrm{Re}_h \cdot \frac{h}{R} \ll 1$, so lubricant density can be ignored. In most cases, environmental pressure effect can also be ignored.

With light loading and above parameters, minimum thickness of lubricant film h_m:

$$h_m = f(R, \ r; \ \mu; \ W, \ v). \tag{6.17}$$

Taking R, W, and v as a unit system generates dimensionless relationship:

$$\frac{h_m}{R} = f\left(\frac{r}{R}, \ \frac{\mu v R}{W}\right), \tag{6.18}$$

where dimensionless parameter $\frac{\mu v R}{W}$ represents ratio of viscosity effect to loading effect.

With heavy loading, lubricant pressure is high enough to make the *viscosity coefficient* of liquid lubricant *increase with pressure* and an empirical power law describes pressure effect:

$$\mu = \mu_0 \cdot e^{\alpha(p - p_a)}, \tag{6.19}$$

where $\mu_0 = $ viscosity coefficient at atmospheric pressure p_a and empirical constant $\alpha = $ *pressure coefficient*. In addition, elastic deformation of axles and bearings should be considered with *Young's modulus* and Poisson's ratio used as governing parameters.

In comparison with cases of light loading, heavy loading has three additional parameters: pressure coefficient for viscosity and two elastic constants. Minimum thickness of lubricant film h_m:

$$h_m = f(R, \ r; \ \mu_0 \ \alpha; \ W, \ v; \ E, \ v). \tag{6.20}$$

Using R, W and v as a unit system generates:

$$\frac{h_m}{R} = f\left(\frac{r}{R}, \ \frac{\mu_0 v R}{W}, \ \frac{\alpha W}{R^2}, \ \frac{ER^2}{W}, \ v\right). \tag{6.21}$$

$\frac{\mu_0 v}{ER}$ and αE can replace $\frac{\mu_0 v R}{W}$ and $\frac{\alpha W}{R^2}$ because:

$$\frac{\mu_0 v}{ER} = \frac{\mu_0 v R}{W^2} \left/ \frac{ER^2}{W} \right. \quad \text{and} \ \alpha E = \frac{\alpha W}{R^2} \cdot \frac{ER^2}{W}.$$

Thus:

$$\frac{h_m}{R} = f\left(\frac{r}{R}, \ \frac{\mu_0 v}{ER}, \ \alpha E, \ \frac{W}{ER^2}, \ v\right), \tag{6.22}$$

where $\frac{\mu_0 v}{ER} = $ ratio of viscosity effect to elasticity effect. The particular form of function f should be determined by experiments.

For geometrically similar model testing and for a certain range of variation for parameters $\frac{\mu_0 v}{ER}$, αE and $\frac{W}{ER^2}$, experimental results can be:

$$\frac{h_m}{R} = c(v) \cdot (\alpha E)^{\xi} \cdot \left(\frac{\mu_0 v}{ER}\right)^{\eta} \cdot \left(\frac{W}{ER^2}\right)^{\zeta}, \tag{6.23}$$

where power constants ξ, η and ζ are determined by experiments.

For gas bearings, compressibility effect is usually considered.

6.3 Fluttering of Airfoils

When water or gas flows past an airfoil or a vessel, fluid motion couples with elastic deformation of the solid body and intensive vibration may occur. In certain conditions, a solid body absorbs energy so continuously from fluid flow that the amplitude of vibration becomes considerable. Vibration of elastic solids is called *fluttering* and in a fluttering problem, viscosity and compressibility are generally insignificant.

The fluttering of an airfoil illustrates the principle for simulating the phenomenon of fluttering (Fig. 6.1).

An airfoil that is disturbed and loses equilibrium produces flexural vibrations and torsional vibrations, due to elastic recoverability. Torsional vibration causes the attack angle of the airfoil to oscillate and results in periodic variation in airfoil lift. Flexural vibration adds vertical velocity that makes the attack angle oscillate and causes intense airfoil vibration. With low air velocity, the airfoil absorbs energy from the air stream and transforms the energy into vibration energy. Vibrations of air stream and airfoil have finite amplitudes. When air velocity is above a certain value, the energy absorbed cannot maintain vibration at finite

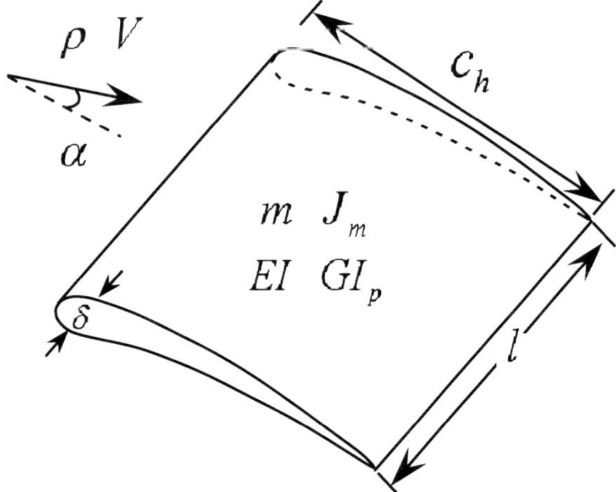

Fig. 6.1 Fluttering of an airfoil

amplitude and amplitude becomes greater and greater. This *critical velocity* is denoted as v_{cr} and called fluttering velocity.

It is assumed that aspect length $= l$, characteristic chord $= c_h$, characteristic thickness $= \delta$, characteristic attack angle $= \alpha$, characteristic mass per unit length of airfoil $= m$, characteristic moment of inertia for unit length of airfoil $= J_m$, characteristic flexural rigidity EI is the product of Young's modulus E and inertial moment I, characteristic torsional rigidity GI_p is the product of *shear modulus G* and polar inertial moment I_p and air density is ρ_0. Critical velocity for fluttering v_{cr}:

$$v_{cr} = f\left(l,\ c_h,\ \delta;\ \alpha;\ m,\ J_m;\ EI,\ GI_p;\ \rho_0\right). \tag{6.24}$$

Dimensions of these variables:

$$[v_{cr}] = LT^{-1}, \qquad [EI] = ML^3T^{-1}, \qquad [GI_p] = ML^3T^{-1},$$

$$[l] = L, \quad [c_h] = L, \quad [\delta] = L, \quad [m] = ML^{-1}, \quad [J_m] = ML,$$

$$[\rho_0] = ML^{-3}, \qquad [\alpha] = M^0L^0T^0.$$

Using dimensionless formulation, critical velocity for fluttering:

$$\frac{v_{cr}}{\left(\frac{EI}{m}\right)^{1/2}\Big/ l} = f\left(\frac{c_h}{l},\ \frac{\delta}{l},\ \alpha;\ \frac{J_m}{mc_h^2},\ \frac{EI}{GI_p},\ \frac{\rho_0 c_h^2}{m}\right). \tag{6.25}$$

All similarity parameters involved in function f have distinct significance: $\frac{c_h}{l}, \frac{\delta}{l}$ and α are geometrical similarity parameters; $\frac{J_m}{mc_h^2}$ represents ratio of rotation inertia to translation motion inertia, both determined by mass distribution in the airfoil; $\frac{EI}{GI_p}$ represents ratio of flexure rigidity to torsional rigidity; $\frac{\rho_0 c_h^2}{m}$ represents ratio of attached mass of fluid to airfoil mass.

6.4 *Vortex-Excited Vibration* in Heat Exchangers

When gas flows past a group of tubes in a heat exchanger, *Kármán vortex streets* and eddy shedding can cause resonance in the heat exchanger. To consider the frequency of the vortex street that occur in gas flow past a single cylinder, it is useful to assume that incoming gas velocity $= v$, gas density $= \rho$, gas viscosity coefficient $= \mu$ and cylinder diameter $= d$. When Reynolds number $\frac{\rho v d}{\mu} > 70$, standing eddies originally attached to the rear of the cylinder are alternately and periodically shed and carried downstream, generating a Kármán vortex street. frequency of vortex street f_K:

$$f_K = f(v,\ \rho,\ \mu,\ d). \tag{6.26}$$

Taking v, ρ and d as a unit system, (6.26) reduces:

$$\frac{f_K d}{v} = f\left(\frac{\rho v d}{\mu}\right), \tag{6.27}$$

where $\frac{\rho v d}{\mu}$ = Reynolds number, so dimensionless frequency is a function of that Reynolds number. When Reynolds number >300, eddying wake becomes turbulent and function f is basically irrelevant to Reynolds number:

$$\frac{f_K d}{v} = \text{constant}. \tag{6.28}$$

In a heat exchanger, gas flows past a group of tubes. Compared with flow past a single cylinder, flow past multiple cylinders has additional governing parameters between two neighboring tubes: horizontal distance l_x and vertical distance l_y. Frequency of eddies shedding into vortex street f_K:

$$f_K = f\left(v,\ \rho,\ \mu,\ d,\ l_x,\ l_y\right). \tag{6.29}$$

Corresponding dimensionless eddy–shedding frequency:

$$\frac{f_K d}{v} = f\left(\frac{\rho v d}{\mu},\ \frac{l_x}{d},\ \frac{l_y}{d}\right). \tag{6.30}$$

Similar to the single cylinder case, when Reynolds number >300, eddying wake becomes turbulent and function f is basically irrelevant to Reynolds number:

$$\frac{f_K d}{v} = f\left(\frac{l_x}{d},\ \frac{l_y}{d}\right). \tag{6.31}$$

For a group of tubes having given geometrical arrangement, a combination of $\left(\frac{l_x}{d},\ \frac{l_y}{d}\right)$ is constant and $\frac{f_K d}{v}$ approximately equals constant c that can be determined by experiments. Thus, eddy–shedding frequency:

$$f_K = c \cdot \frac{v}{d}. \tag{6.32}$$

Intense noise occurs if shedding frequency f_K nears inherent frequency f_c of the heat exchanger's gas chamber. Intense vibration occurs if f_K nears inherent frequency f_s of the heat exchanger's structure. In engineering design, frequencies f_K, f_c and f_s should be separated to prevent resonance.

Chapter 7
Hydro-Elasto-Plastic Modeling

A hydro-elasto-plastic model shows fluid-like behavior and solid-like behavior of a body subjected to intense dynamic loading and deformation. Governing parameters and similarity parameters useful in solving real problems of explosion and high velocity impact are given.

7.1 Hydro-Elasto-Plastic Model

In continuum mechanics, a *hydro-elasto-plastic model* is a material model in which a body is subjected to such intense dynamic loading and deformation that, in order to describe its motion and deformation, the body must be regarded as behaving like a fluid and as behaving like a solid having elastic and plastic properties. Such conditions can occur in a body subject to strong explosion action or high velocity impact.

Problems arose in the 1960s in connection with the study of *underground nuclear explosions. C. M. Cheng* and *B. M. Xie* [1] devised a constitutive relation that permits rock mass to flow like fluid under pressure far exceeding strength of the rock and that permits rock mass to deform like an elasto-plastic solid under pressure comparable to rock strength. Since then, hydro-elasto-plastic modeling has become important in studying underground nuclear explosions, anti-missile systems in space, armor penetration, explosive welding and explosions that are introduced into the earth by deep drilling techniques.

In a problem concerning explosion and high velocity impact, materials such as rock or metals are conventionally regarded as being solids. However, such materials are subject to intense dynamic loading that causes their behavior to change. Early in the loading process and in the region neighboring the loading source, material yield strength is much lower than characteristic pressure (either static pressure or dynamic pressure) and the material behaves like fluid. Later in the loading process and in the region far from the loading source, pressure becomes lower and material behaves like an elasto-plastic solid.

Q.-M. Tan, *Dimensional Analysis*, DOI: 10.1007/978-3-642-19234-0_7,
© Springer-Verlag Berlin Heidelberg 2011

Table 7.1 lists main parameters for several typical applications of a hydro-elasto-plastic model. Y = yield limit, p = characteristic pressure (either static pressure or dynamic pressure) and magnitudes of ratio $\frac{Y}{p}$ are vital parameters that reflects the essentials of actual problems.

If a high velocity jet generated by a shaped charge causes armor penetration and if $\frac{Y}{p} < 10^{-2}$, jet and armor can both be regarded as fluids. Likewise, in the early stage of underground nuclear explosion or before a shock wave leaves a remote explosion source, ratio $\frac{Y}{p}$ is so small that rock can be regarded as a fluid. Rock behaves as a solid in the latter stage of such explosion or in a region remote from the explosion source.

Hydro-elasto-plastic bodies have motion peculiarities that differ from those of fluids and elasto-plastic bodies:

1. Action of plastic stress increases temperature of medium, pressure and density. Such changes cause coupling of shear deformation and volumetric deformation and cause coupling of mechanical effect and thermal effect;
2. Generally, variations in temperature, pressure and strain rate affect yield limit.

In hydro-elasto-plastic modeling, fluid properties (reflecting volumetric deformation) and solid properties (mainly reflecting distortion) couple together and three fundamental assumptions can be made:

Assumption 1. Stress σ_{ij} and strain e_{ij} are composed of volumetric deformation part and distortion part:

Table 7.1 Hydro-elasto-plastic model

Values of Y, p, and Y/p in typical cases

Type	Y (MPa)	p (MPa)	Y/p	Remarks
Underground nuclear explosion	5×10^2	10^6	5×10^{-4}	p-initial pressure
Blasting in rock and soil	$10 \sim 5 \times 10^2$	5×10^3	$2 \times 10^{-3} \sim 10^{-1}$	p-initial pressure
Explosive hardening	10^3	10^4	10^{-1}	p-detonation pressure
Impact of meteorites	$10^2 \sim 5 \times 10^2$	$10^5 \sim 10^6$	$10^{-3} \sim 5 \times 10^{-4}$	p-initial pressure at shock front: ρUu U- shock front velocity, u-impact velocity
Shaped charge	10^3	$10^3 \sim 7 \times 10^4$	$1 \sim 1.4 \times 10^{-2}$	p-characteristic dynamic pressure
Projectile penetration	10^3	5×10^3	2×10^{-1}	p-characteristic dynamic pressure
Explosive welding	10^3	$10^4 \sim 5 \times 10^4$	$10^{-1} \sim 2 \times 10^{-2}$	p-characteristic dynamic pressure

$$\sigma_{ij} = \frac{\sigma_{ii}}{3} \cdot \delta_{ij} + s_{ij} \quad \text{and} \quad e_{ij} = \frac{e_{ii}}{3} \cdot \delta_{ij} + \varepsilon_{ij},$$

where $\frac{\sigma_{ii}}{3}$ can be regarded as negative pressure or pressure $p = -\frac{\sigma_{ii}}{3}$, and e_{ii} can be regarded as relative variation of volume $\frac{\Delta V}{V}$ or $\frac{\Delta V}{V} = e_{ii}$, so:

$$\sigma_{ij} = -p \cdot \delta_{ij} + s_{ij}, \quad \text{and} \quad \frac{de_{ij}}{dt} = \frac{1}{3V}\frac{dV}{dt} \cdot \delta_{ij} + \frac{d\varepsilon_{ij}}{dt} \tag{7.1}$$

where s_{ij} = deviation stress, ε_{ij} = deviation strain, and $\frac{d\varepsilon_{ij}}{dt}$ = deviation strain rate.

Assumption 2. Strain e_{ij} is composed of elastic component e_{ij}^e and plastic component e_{ij}^p, so:

$$e_{ij} = e_{ij}^e + e_{ij}^p = \left(\frac{e_{ii}}{3} \cdot \delta_{ij} + \varepsilon_{ij}^e\right) + \varepsilon_{ij}^p, \tag{7.2}$$

where superscript "e" = elastic component and superscript "p" = plastic component. Because plastic volumetric deformation in metallic materials is near zero (unless rock or soil has shear-dilatation), it is possible to assume that elastic strain rate $= \frac{de_{ij}^e}{dt} = \frac{1}{3V}\frac{dV}{dt} \cdot \delta_{ij} + \frac{d\varepsilon_{ij}^e}{dt}$ and plastic strain rate $= \frac{de_{ij}^p}{dt} = \frac{d\varepsilon_{ij}^p}{dt}$.

Assumption 3. Constitutive relationship has components that correspond to fluid and solid.

The Mie–Grüneisen equation of state commonly describes properties of fluids:

$$p - p_k(V) = \frac{\gamma}{V} \cdot [E - E_k(V)], \tag{7.3}$$

where p = pressure, V = specific volume, E = specific internal energy, γ = *Grüneisen Coefficient*, $p_k(V)$ = cold pressure, and $E_k(V)$ = cold specific internal energy.

In general, shock compression tests with variable impact intensity can be used to obtain *Hugoniot relationships* relating state variables $p_H(V)$ and $E_H(V)$ behind the shock front. The functions in Hugoniot relationships $p_H(V)$ and $E_H(V)$ can be used to determine $p_k(V)$ and $E_k(V)$. Substituting experimental data for pressure $p_H(V)$ and specific internal energy $E_H(V)$ into equation of state (7.3) generates:

$$p_H - p_k(V) = \frac{\gamma}{V} \cdot [E_H - E_k(V)] \tag{7.4}$$

Subtracting (7.3) and (7.4) produces useful equation of state:

$$p - p_H(V) = \frac{\gamma}{V} \cdot [E - E_H(V)] \tag{7.5}$$

An empirical linear relationship for many materials relates propagation velocity U of the front of intense shock wave and particle velocity u behind that front:

$$U = c + bu, \tag{7.6}$$

where c and b = empirical constants for special material used in shock compression testing. This linear relationship can express particular form of $p_H(V)$ and $E_H(V)$, so γ, c, and b can represent governing parameters that reflect volumetrical deformation properties of material.

Properties of solids can be described by a stress–strain relationship that includes elasticity, yield and plasticity:

1. Hooke's elasticity law:

$$de_{ij} = \frac{1}{2G} \cdot ds_{ij} + \frac{1-2v}{3E} d\sigma_{kk} \cdot \delta_{ij} \tag{7.7}$$

where E and v = elastic constants and $G = \frac{E}{2(1+v)}$.

2. *Von Mises yield criterion*:

If deviation stress s_{ij} satisfies yield criterion, plastic yield occurs in material. For metallic materials, von Mises yield criterion is commonly used:

$$J_2 = -\left(s_{11}s_{22} + s_{22}s_{33} + s_{33}s_{11} - s_{11}^2 - s_{22}^2 - s_{33}^2\right) = \frac{Y^2}{3}, \tag{7.8}$$

or

$$\tau_i\left(= J_2^{1/2}\right) = \frac{Y}{3^{1/2}},$$

where J_2 = second invariant of s_{ij} and yield strength Y is expressed as a function of strain rate, pressure, and temperature.

3. *Prandtl–Reuss relationship*

The *Prandtl–Reuss relationship* describes plastic behavior after yielding:

$$de_{ij} = \frac{1}{2G} \cdot ds_{ij} + \frac{1-2v}{3E} d\sigma_{kk} \cdot \delta_{ij} + \left(\frac{3}{2} de_{klp} de_{klp}\right)^{1/2} \cdot \frac{s_{ij}}{Y}. \tag{7.9}$$

Governing parameters that reflect properties of elasto-plastic deformation can be represented by E, v, and Y.

Ultimately, six material parameters can be used in hydro-elasto-plastic modeling:

$$\gamma, c, b \quad \text{and } E, v, Y. \tag{7.10}$$

7.2 Similarity Parameters in Problems Related to Chemical Explosions

In dealing with chemical explosions, explosives (denoted by subscript "e") and the object affected by the explosion (e.g., working object or target denoted by subscript "t") generate governing parameters that include:

1. Parameters for explosive: explosive mass Q, loading density ρ_e, chemical energy released by unit mass of explosive E_e, dilatation index of the explosion product γ_e;
2. Initial parameters for object affected by explosion: characteristic length l_t, initial density ρ_t, initial pressure p_t;
3. Constitutive parameters for object of explosion: properties of fluid behavior (Grüneisen coefficient v_t, *Hugoniot relationship* constants c_t and b_t) and properties of solid behavior (elastic constants E_t and v_t and yield limit Y_t).

Taking l_t, ρ_t, and Y_t as a unit system produces dimensionless parameters:

1. Parameters for explosive: $\frac{Q}{\rho_e l_t^3}$, $\frac{\rho_e E_e}{Y_t}$, γ_e;
2. Initial parameters for object of explosion: $\frac{\rho_t}{\rho_e}$, $\frac{p_t}{Y_t}$;
3. Constitutive parameters for object of explosion: γ_t, $\frac{\rho_t c_t^2}{Y_t}$, b_t, $\frac{E_t}{Y_t}$, and v_t,

where relative mass of explosive $\frac{Q}{\rho_e l_t^3}$ can sometimes be replaced by geometrical similarity parameter $\frac{(Q/\rho_e)^{1/3}}{l_t}$.

Some parameters can be ignored in certain realistic problems. For example, if initial pressure of object of explosion $p_t \ll$ object strength Y_t, parameter $\frac{p_t}{Y_t}$ can be ignored. As another example, elastic parameters $\frac{E_t}{Y_t}$ and v_t can be ignored if elastic deformation $\ll$ plastic deformation.

7.3 Similarity Parameters in Problems Related to High Velocity Impact

In problems of high velocity impact, governing parameters are generated from a high velocity impactor (simply called projectile and denoted by subscript "p") and the object of impact (e.g., working object or target denoted by subscript "t"). These governing parameters include:

1. Initial parameters for projectile: characteristic length l_p, characteristic velocity v_p, initial density ρ_p and initial pressure p_p;
2. Constitutive parameters for projectile: Grüneisen Coefficient γ_p, constants in the *Hugoniot relationship* c_p and b_p; elastic constants E_p and v_p and yield limit Y_p.

3. Initial parameters for target: characteristic length l, initial density ρ_t and initial pressure p_t;
4. Constitutive parameters for target: Grüneisen coefficient γ_t, constants in the *Hugoniot relationship* c_t and b_t; elastic constants E_t and v_t and yield limit Y_t.

Taking l_t, v_p and Y_t as a unit system produces dimensionless parameters:

1. Initial parameters for projectile: $\frac{l_p}{l_t}$, $\frac{v_p}{\sqrt{Y_t/\rho_p}}$, and $\frac{p_p}{Y_t}$;
2. Constitutive parameters for projectile: γ_p, $\frac{c_p}{v_p}$, b_p, $\frac{E_p}{Y_t}$, and v_p;
3. Initial parameters for target: $\frac{p_t}{Y_t}$;
4. Constitutive parameters for target: γ_t, $\frac{c_t}{v_p}$, b_t, $\frac{E_t}{Y_t}$, and v_t.

If similar materials are used in model and prototype and if effects of initial pressure and elastic deformation are insignificant, dimensionless parameters reduce:

$$\frac{l_p}{l_t}, \ \frac{v_p}{\sqrt{Y_t/\rho_p}}, \ \frac{v_p}{c_t}, \ \gamma_t \quad \text{and } b_t,$$

where $\frac{v_p}{\sqrt{Y_t/\rho_p}}$ = square root of $\frac{\rho_p v_p^2}{Y_t}$, $\frac{\rho_p v_p^2}{Y_t}$ is commonly called damage number and represents ratio of inertia effect to *strength effect*, $\frac{v_p}{c_t}\left(=\frac{v_p}{c_p}\cdot\frac{c_p}{c_t}=M_p\cdot\frac{c_p}{c_t}\right)$ is a equivalent *Mach number* and represents ratio of inertia effect to *compressibility effect*.

Reference

1. Cheng, C.M., Xie, B.M.: A Proposal for Computational Model of Underground Explosion. Research Report, Institute of Mechanics, CAS. May, 1965. (in Chinese)

Chapter 8
Similarity Laws for Explosions

This chapter discusses propagation laws of explosion waves and fundamental principles and similarity laws of explosions in engineering technology applications that mainly concern explosive forming, explosive welding and blasting.

Explosion is a phenomenon that occurs quickly in confined space, transforms one type of energy into another type and is accompanied by high pressure, high temperature and *shock waves*. Chemical explosions transform chemical energy into mechanical energy. Nuclear explosions transform nuclear energy into mechanical energy. Electrical explosions transform electrical energy into mechanical energy. Broadly speaking, high velocity impact is also a kind of explosion phenomenon because it rapidly transforms a type of mechanical energy (kinetic energy) into another type of mechanical energy (deformation energy) and thermal energy. Such energy transformation occurs quickly in confined space, so power and power density are considerable.

Generally, nuclear explosion intensity is measured by *TNT yield*. Chemical energy released from one kilogram of TNT is about 1,000 kcal, power is about 5.6×10^8 Kwatt and power density is about 9.3×10^5 Kwatt/cm^3. To put these numbers in perspective, a modern turbo-generator has maximum power of about 1.3×10^6 Kwatt but generator volume far exceeds the volume of one kilogram of TNT. The first atomic bomb exploded in New Mexico in 1945 yielded the equivalent of about 17 Kton of TNT, but that bomb was much smaller than the generator in terms of volume and much higher than the generator in terms of power and power density.

When high power and high power density abruptly release high pressure, a strong *shock wave* affects surrounding media (air, liquid or solid) and there is serious deformation or damage. This process is often applied to research and development on ordnance and mechanical working. High velocity impact and its effects are discussed in detail in Chap. 9.

Q.-M. Tan, *Dimensional Analysis*, DOI: 10.1007/978-3-642-19234-0_8,
© Springer-Verlag Berlin Heidelberg 2011

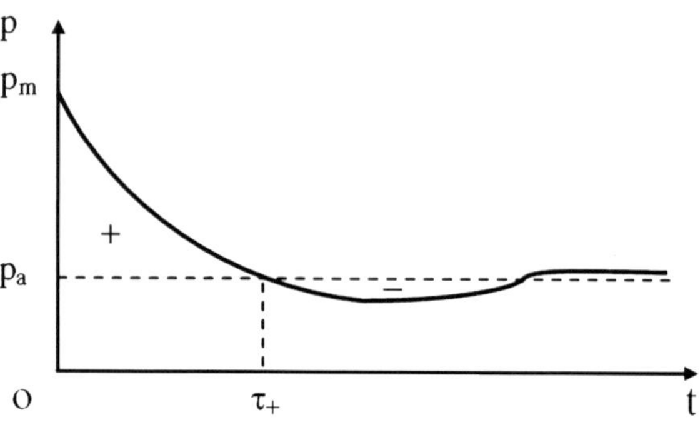

Fig. 8.1 Pressure variation of explosion wave in air

8.1 Explosion Waves in Air and Water

8.1.1 Explosion Waves in Air

After detonation of explosive charges, artillery shells or bombs in air, *shock waves* form and propagate into surrounding space. The shock wave front abruptly increases air pressure and simultaneously, particle velocity increases sharply. Figure 8.1 sketches pressure variation at distance R from an explosion source. When the shock wave front arrives, pressure p instantly jumps from undisturbed atmospheric pressure p_a to maximum value p_m (peak pressure). Thereafter, pressure decreases continuously and, with time lapse τ_+, pressure returns to undisturbed pressure p_a. The whole process divides into phases of *positive overpressure* and *negative overpressure*, based on overpressure $p - p_a$ being positive or negative (Fig. 8.1). A solid body's reaction to a shock wave results in deformation or damage that mainly depend on positive overpressure, particularly *peak* overpressure $(p - p_a)_m$, positive overpressure duration τ_+ and positive overpressure impulse $I = \int_0^{\tau_+} (p - p_a)\mathrm{d}t$.

Exploding a spherical charge is the simplest and most common case of explosion. After explosion, a spherical shock wave forms and propagates outward. Governing parameters determining the intensity of shock wave relate to three factors:

1. Explosive: mass of charge Q, loading density ρ_e, chemical energy released per unit mass of explosive E_e and dilatation index of detonation products γ_e;
2. Air: initial pressure p_a, initial density ρ_a and adiabatic index γ_a;
3. Distance from center of explosive charge: R.

Peak overpressure $(p - p_a)_m$ and duration of positive overpressure τ_+ are functions of parameters that are associated with these factors:

$$\begin{cases} (p - p_a)_m = f(Q,\ \rho_e,\ E_e, \gamma_e;\ p_a,\ \rho_a,\ \gamma_a;\ R), \\ \tau_+ = g(Q,\ \rho_e,\ E_e, \gamma_e; p_a,\ \rho_a,\ \gamma_a;\ R). \end{cases} \tag{8.1}$$

Taking Q, ρ_e and E_e as a unit system reduces (8.1) to dimensionless relationships:

$$
\begin{cases}
\dfrac{(p-p_a)_m}{\rho_e E_e} = f\left(\gamma_e;\ \dfrac{p_a}{\rho_e E_e},\ \dfrac{\rho_a}{\rho_e},\ \gamma_e;\ \dfrac{R}{(Q/\rho_e)^{1/3}}\right), \\[4mm]
\dfrac{\tau_+}{(Q/\rho_e)^{1/3}\big/ E_e^{1/2}} = g\left(\gamma_e;\ \dfrac{p_a}{\rho_e E_e},\ \dfrac{\rho_a}{\rho_e},\ \gamma_e;\ \dfrac{R}{(Q/\rho_e)^{1/3}}\right).
\end{cases}
\tag{8.2}
$$

If explosive used in the model is the same as that used in the prototype and modeling is in air, six parameters are constant:

$$
\left(\rho_e,\ E_e,\ \gamma_e;\ p_a,\ \rho_a,\ \gamma_a\right) = \text{constant.}
$$

Thus, dimensionless peak overpressure and duration of positive overpressure reduces:

$$
\begin{cases}
\dfrac{(p-p_a)_m}{\rho_e E_e} = f\left(\dfrac{R}{(Q/\rho_e)^{1/3}}\right), \\[4mm]
\dfrac{\tau_+}{(Q/\rho_e)^{1/3}\big/ E_e^{1/2}} = g\left(\dfrac{R}{(Q/\rho_e)^{1/3}}\right).
\end{cases}
\tag{8.3a}
$$

or

$$
\begin{cases}
(p-p_a)_m = \rho_e E_e \cdot f\left(\dfrac{R}{(Q/\rho_e)^{1/3}}\right), \\[4mm]
\tau_+ = (Q/\rho_e)^{1/3}\big/ E_e^{1/2} \cdot g\left(\dfrac{R}{(Q/\rho_e)^{1/3}}\right).
\end{cases}
\tag{8.3b}
$$

$(Q/\rho_e)^{1/3} =$ characteristic length of explosive charge, so peak overpressure and duration of positive overpressure follow geometrical similarity laws. The following expressions are commonly used in engineering:

$$
\begin{cases}
(p-p_a)_m = f\left(\dfrac{R}{Q^{1/3}}\right), \\[4mm]
\tau_+ = Q^{1/3} \cdot g\left(\dfrac{R}{Q^{1/3}}\right).
\end{cases}
\tag{8.4}
$$

Dimensions in both sides of expressions in (8.4) differ, so those dimensional expressions for $(p-p_a)_m$ and τ_+ are unreasonable rigorously. However, using these expressions in engineering is based on two prerequisites that must be kept in mind:

Prerequisite 1. The same type of explosive is used in model and prototype, so ρ_e, E_e and γ_e are constant and their influences on $(p - p_a)_m$ and τ_+ are also constant.

Prerequisite 2. Units for all variables in expressions are given, including four independent variables ρ_e, E_e, R and Q and two dependent variables $(p - p_a)_m$ and τ_+.

Due to these prerequisites, expressions (8.4) are relationships between pure numbers and variables are magnitudes of corresponding quantities.

Systematic experiments produce empirical formulae for spherical charges of different types of explosives. For example, if m, kg and s are used for units of length, mass and time, respectively, kgf/cm^2 represents unit of overpressure and relative distance R_* is defined:

$$R_* = \frac{R(\text{m})}{(Q(\text{kg}))^{1/3}}, \tag{8.5}$$

where m and kg represent units of length and mass and R_* is regarded as a pure number, experimental results for TNT in the range of $1 \leq R_* \leq 15$ can be processed:

$$\begin{cases} (p - p_a)_m = \dfrac{0.84}{R_*} + \dfrac{0.7}{R_*^2} + \dfrac{7.4}{R_*^3} \\[2mm] \tau_+ = 1.35 \times 10^{-3} \cdot R_*^{1/2}. \end{cases} \tag{8.6}$$

8.1.2 Explosion Waves in Water

Explosion waves in water and explosion waves in air have many similar characteristics but there are two major differences:

Difference 1. It is more difficult to compress water than to compress air, but the state of equation of water can be approximated:

$$p - p_a = B\left[\left(\frac{\rho}{\rho_w}\right)^{\gamma_w} - 1\right], \tag{8.7}$$

where ρ_w = density of water under normal pressure, pressure coefficient $B = 3050$ kgf/cm^2 and adiabatic index $\gamma_w = 7.15$. Roughly speaking, shock waves with peak pressure < 1000 kgf/cm^2 can be regarded as being acoustic waves in water.

Difference 2. In explosion in air, environmental pressure or undisturbed pressure p_a cannot be ignored. In explosion in water, environmental pressure or undisturbed pressure p_a can generally be ignored. There is no need to introduce overpressure whether positive overpressure or negative overpressure.

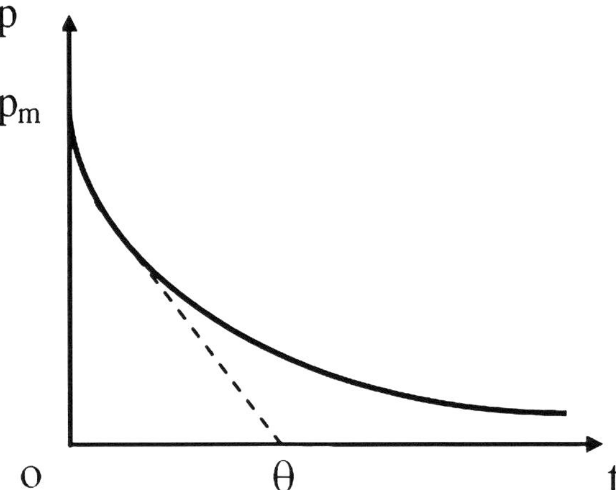

Fig. 8.2 Pressure variation of explosion wave in water

Experiments on shock waves induced by explosion in water show that variation of pressure with time can be approximated by using exponential decay law:

$$p = p_m e^{-t/\theta}, \tag{8.8}$$

where $p_m = $ *peak pressure* and $\theta = $ *duration of pressure*. Figure 8.2 sketches pressure variation with time related to explosions in water.

Three factors affect governing parameters that determine shock wave intensity:

1. Explosive: mass of charge Q, loading density ρ_e, chemical energy released per unit mass of explosive E_e and dilatation index γ_e;
2. Water: initial density ρ_w, pressure coefficient B and adiabatic index γ_w;
3. Distance from center of explosive charge: R.

Peak pressure p_m and duration of pressure θ are functions of parameters related to these three factors:

$$\begin{cases} p_m = f(Q, \ \rho_e, \ E_e, \ \gamma_e; \ \rho_w, \ B, \ \gamma_w; \ R), \\ \theta = g(Q, \ \rho_e, \ E_e, \ \gamma_e; \ \rho_w, \ B, \ \gamma_w; \ R). \end{cases} \tag{8.9}$$

Taking Q, ρ_e and E_e as a unit system reduces (8.9) to dimensionless relationship:

$$\begin{cases} \dfrac{p_m}{\rho_e E_e} = f\left(\gamma_e; \ \dfrac{\rho_w}{\rho_e}, \ \dfrac{B}{\rho_e E_e}, \ \gamma_w; \ \dfrac{R}{(Q/\rho_e)^{1/3}} \right), \\[4mm] \dfrac{\theta}{(Q/\rho_e)^{1/3} \big/ E_e^{1/2}} = g\left(\gamma_e; \ \dfrac{\rho_w}{\rho_e}, \ \dfrac{B}{\rho_e E_e}, \ \gamma_w; \ \dfrac{R}{(Q/\rho_e)^{1/3}} \right). \end{cases} \tag{8.10}$$

If explosive used in the model is the same as that used in the prototype and modeling is in water, (8.10) reduces:

$$
\left\{
\begin{aligned}
\frac{p_m}{\rho_e E_e} &= f\left(\frac{R}{(Q/\rho_e)^{1/3}}\right), \\[2ex]
\frac{\theta}{(Q/\rho_e)^{1/3}\big/ E_e^{1/2}} &= g\left(\frac{R}{(Q/\rho_e)^{1/3}}\right).
\end{aligned}
\right.
\tag{8.11a}
$$

or

$$
\left\{
\begin{aligned}
p_m &= \rho_e E_e \cdot f\left(\frac{R}{(Q/\rho_e)^{1/3}}\right), \\[2ex]
\theta &= \frac{(Q/\rho_e)^{1/3}}{E_e^{1/2}} \cdot g\left(\frac{R}{(Q/\rho_e)^{1/3}}\right).
\end{aligned}
\right.
\tag{8.11b}
$$

Peak pressure and duration of pressure for explosion in water follow *geometrical similarity laws* because $(Q/\rho_e)^{1/3}$ represents characteristic length of explosive charge. Simplified expressions commonly used in engineering are:

$$
\left\{
\begin{aligned}
p_m &= f\left(\frac{R}{Q^{1/3}}\right), \\[2ex]
\theta &= Q^{1/3} \cdot g\left(\frac{R}{Q^{1/3}}\right).
\end{aligned}
\right.
\tag{8.12}
$$

Expression (8.12) applies only to type of explosives prescribed and prescribed units for variables used.

For TNT with loading 1.52 g/cm^3, if m, kg, kgf/cm^2 and ms are units of length, mass, pressure and time, respectively, experimental data concerning magnitudes of R, Q, p_m and θ can be processed by empirical relationships:

$$
\left\{
\begin{aligned}
p_m &= a \cdot \left(\frac{R}{Q^{1/3}}\right)^{\alpha}, \\[2ex]
\theta &= b \cdot Q^{1/3} \cdot \left(\frac{R}{Q^{1/3}}\right)^{\beta}.
\end{aligned}
\right.
\tag{8.13a}
$$

where a, α and b, β are pure numbers:

$$
\left\{
\begin{aligned}
a &= 534, & \alpha &= -1.13; \\
b &= 0.110, & \beta &= 0.24.
\end{aligned}
\right.
\tag{8.13b}
$$

8.1.3 *Intense Explosion of Point Source*

In 1941, Taylor [1] studied energy release in the nuclear fission weapons (or atomic bombs) that were crucial during World War II. Taylor ingeniously applied

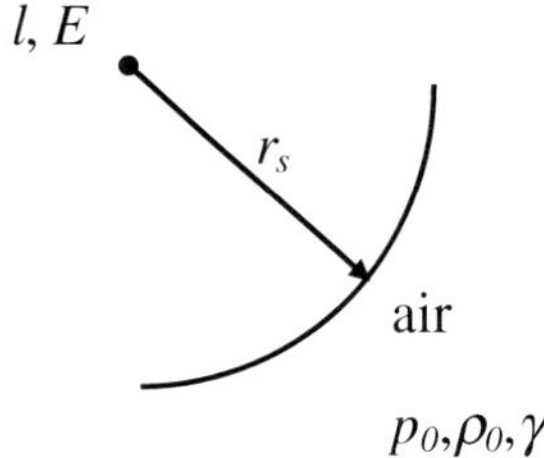

Fig. 8.3 Spherical explosion wave for atomic bomb

dimensional analysis in his theory and found that when explosion waves in air are sufficiently intense, explosion source size and environmental pressure can be ignored and such intense explosion can be viewed as a burst of extremely high energy from a point source. Taylor's theory of intensive explosion from a point source and propagation laws of explosion waves allowed derivation of a self-similar solution for a flow field caused by intense explosion of point source. von Neumann [2] improved Taylor's calculations and obtained a close form of analytical solution. In 1946, Sedov [3] independently obtained the same solution for the problem of point explosion. The propagation law of explosion waves and self-similar solution of the point explosion flow field deserve detailed consideration.

1. Propagation law of explosion waves

It is assumed that an atomic bomb with characteristic length l releases an extremely large amount of energy E in a very short time. Thereafter, surrounding air compresses abruptly and an intensive shock wave forms and propagates outward (Fig. 8.3). If it is assumed that surrounding air has initial pressure p_0 and density ρ_0 and compressibility of air is represented by adiabatic index γ and if the explosion center of the atomic bomb is regarded as the origin of polar coordinates, the explosion wave front can be approximated as a spherical surface with a radius that increases with time. The radius of front r_s is a function of time t and related governing parameters are l, E, p_0, ρ_0 and γ:

$$r_s = f(t; \ l, \ E; \ p_0, \ \rho_0; \ \gamma). \tag{8.14}$$

Taking l, E and ρ_0 as a unit system reduces (8.14) to dimensionless relationship:

$$\frac{r_s}{(E/\rho_0)^{1/5} \cdot t^{2/5}} = f\left(\frac{l}{(E/\rho_0)^{1/5} \cdot t^{2/5}}, \ \frac{p_0}{E^{2/5} \rho_0^{3/5} t^{-6/5}}, \ \gamma\right). \tag{8.15}$$

The first two dimensionless independent variables in (8.15) $\frac{l}{(E/\rho_0)^{1/5} \cdot t^{2/5}}$, $\frac{p_0}{E^{2/5} \rho_0^{3/5} t^{-6/5}}$ can transform into $\frac{l}{r_s}$ and $\frac{p_0}{E/r_s^3}$ if factor $\frac{r_s}{(E/\rho_0)^{1/5} \cdot t^{2/5}}$ is introduced into the transformation:

$$\frac{l}{r_s} = \frac{l}{(E/\rho_0)^{1/5} \cdot t^{2/5}} \bigg/ \frac{r_s}{(E/\rho_0)^{1/5} \cdot t^{2/5}}$$

and

$$\frac{p_0}{E/r_s^3} = \frac{p_0}{E^{2/5}\rho_0^{3/5}t^{-6/5}} \cdot \left[\frac{r_s}{(E/\rho_0)^{1/5}t^{2/5}}\right]^3 \cdot$$

The result is a new dimensionless relationship for $\frac{r_s}{(E/\rho_0)^{1/5}\cdot t^{2/5}}$:

$$\frac{r_s}{(E/\rho_0)^{1/5}\cdot t^{2/5}} = f\left(\frac{l}{r_s}, \frac{p_0}{E/r_s^3}, \gamma\right). \tag{8.16}$$

In this new relationship, the first independent variable $\frac{l}{r_s}$ represents the ratio of the size of the atomic bomb to radius of the shock front and the second independent variable $\frac{p_0}{E/r_s^3}$ represents the ratio of counter-pressure ahead of the shock wave to the mean pressure of air compressed by that shock wave. If there is concern that propagation distance considerably exceeds radius of atomic bomb l and shock wave intensity considerably exceeds counter-pressure p_0, shock wave intensity decreases during this stage, but $\frac{l}{r_s} \ll 1$ and $\frac{p_0}{E/r_s^3} \ll 1$. These two dimensionless variables $\frac{l}{r_s}$ and $\frac{p_0}{E/r_s^3}$ do not basically influence dimensionless propagation distance $\frac{r_s}{(E/\rho_0)^{1/5}t^{2/5}}$. Thus, l and p_0 are not governing parameters during this stage and propagation distance of shock wave front r_s is considered to be a function of t, E, ρ_0 and γ:

$$r_s = f(t;\ E,\ \rho_0,\ \gamma). \tag{8.17}$$

Taking l, E and ρ_0 as a unit system produces:

$$\frac{r_s}{(E/\rho_0)^{1/5}t^{2/5}} = f(\gamma). \tag{8.18}$$

The right hand side of (8.18) is a function of γ, so $f(\gamma)$ is a constant denoted by ξ_s for given γ. Thus:

$$\frac{r_s}{(E/\rho_0)^{1/5}t^{2/5}} = \xi_s,$$

or

$$r_s(t) = \xi_s \cdot (E/\rho_0)^{1/5}t^{2/5}. \tag{8.19}$$

Expression (8.19) shows that the radius of the explosion wave front expands outward with time to the power of 2/5.

J.E Mack's photographs of fireballs from the 1945 atomic explosion in New Mexico appeared in 1950 and that observational data validated Taylor's 1941 analysis. Taylor used a c.g.s. system to measure variables and take logarithms from (8.19):

Fig. 8.4 $(5/2)\log r_s$ versus $\log t$ for the shock wave front

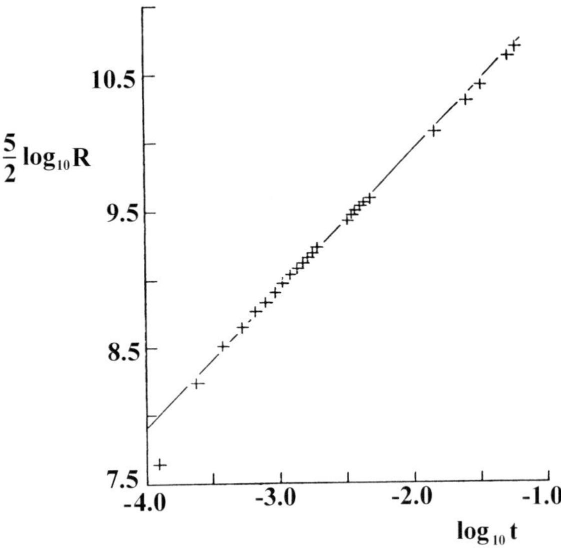

$$\log_{10} r_s = \log_{10} \xi_s + \frac{1}{5}\log_{10}\frac{E}{\rho_0} + \frac{2}{5}\log_{10} t, \tag{8.20a}$$

or

$$\frac{5}{2}\log_{10} r_s - \log_{10} t = \frac{5}{2}\log_{10} \xi_s + \frac{1}{2}\log_{10}\frac{E}{\rho_0}. \tag{8.20b}$$

A diagram with double-logarithm coordinates (8.20b) describes a straight line and, compared with observation data taken from Mack's photographs of the fireballs, theoretical prediction agrees well with observation (Fig. 8.4). The straight line representing observation data points "+":

$$\frac{5}{2}\log_{10} r_s - \log_{10} t = 11.915. \tag{8.21}$$

By analyzing flow field, Taylor found value $\xi_s = 1.033$. Thus:

$$\frac{1}{2}\log\frac{E}{\rho_0} = 11.915 - \frac{5}{2}\log 1.033 = 11.880. \tag{8.22}$$

Because air density $\rho_0 = 1.25 \times 10^{-3}$ g/cm^3, *TNT yield* of the first atomic bomb (1945) was predictable:

$$E = 7.19 \times 10^{20}\text{erg} \approx 17\,\text{Kton TNT yield}. \tag{8.23}$$

When Taylor published this result, the information department of the United States was quite disturbed.

Similar dimensional analysis can be used to derive peak pressure at explosion wave front:

$$p_s = g(\gamma) \cdot \rho_0 (E/\rho_0)^{2/5} t^{-6/5}, \tag{8.24a}$$

or

$$p_s = g(\gamma) \cdot \rho_0 (r_s/t)^2, \tag{8.24b}$$

where $g(\gamma) = $ constant depending on γ.

2. Self-similar solution of explosion flow field

After atomic bomb explosion, a spherical intensive shock wave propagates outward. Air affected by the shock front compresses, pressure and density increase abruptly and air particles behind the shock front accelerate. A flow field that varies with time generates behind the shock front. In this field, pressure p, density ρ and particle velocity v are functions of distance r from explosion center and time t starts from ignition. Assuming intense explosion from source point, atomic bomb size and counter-pressure can be ignored. Thus, the flow field behind the shock wave front is:

$$\begin{cases} p = f_p(r, t; E; \rho_0; \gamma), \\ \rho = f_\rho(r, t; E; \rho_0; \gamma), \\ v = f_v(r, t; E; \rho_0; \gamma). \end{cases} \tag{8.25}$$

Taking l, E and ρ_0 as a unit system produces:

$$\begin{cases} \dfrac{p}{\rho_0^{3/5} E^{2/5} t^{-6/5}} = f_p\left(\dfrac{r}{(E/\rho_0)^{1/5} t^{2/5}}, \ \gamma \right), \\ \dfrac{\rho}{\rho_0} = f_\rho\left(\dfrac{r}{(E/\rho_0)^{1/5} t^{2/5}}, \ \gamma \right) \\ \dfrac{v}{(E/\rho_0)^{1/5} t^{-3/5}} = f_v\left(\dfrac{r}{(E/\rho_0)^{1/5} t^{2/5}} \right). \end{cases} \tag{8.26}$$

Dimensionless independent variables and dependent variables in (8.26):

$$\begin{cases} \xi = \dfrac{r}{(E/\rho_0)^{1/5} t^{2/5}}, \\ p_* = \dfrac{p}{\rho_0^{3/5} E^{2/5} t^{-6/5}}, \\ \rho_* = \dfrac{\rho}{\rho_0}, \\ v_* = \dfrac{v}{(E/\rho_0)^{1/5} t^{-3/5}}. \end{cases} \tag{8.27}$$

Fig. 8.5 Contours for ξ in self-similar solution

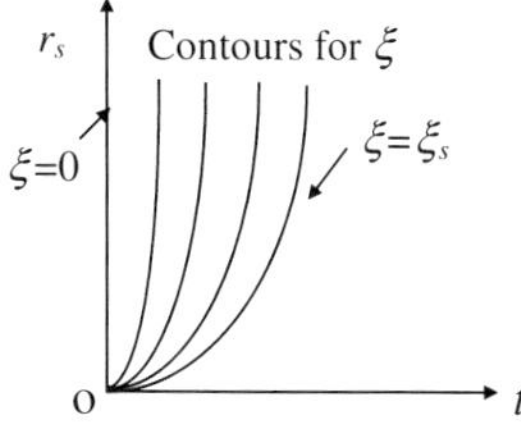

Therefore, dimensionless flow field is:

$$\begin{cases} p_* = f_p(\xi,\ \gamma), \\ \rho_* = f_\rho(\xi,\ \gamma), \\ v_* = f_v(\xi,\ \gamma). \end{cases} \tag{8.28}$$

In the plane with abscissa r and ordinate t, a group of contours of ξ can be drawn (Fig. 8.5):

$$r = \xi \cdot (E/\rho_0)^{1/5} t^{2/5}. \tag{8.29}$$

The flow field behind the shock wave front is located between shock wave front $\xi = \xi_s$ and explosion center $\xi = 0$:

Outer boundary:

$$\xi = \xi_s \quad \text{or } r = \xi_s \cdot (E/\rho_0)^{1/5} t^{2/5} \tag{8.30a}$$

Inner boundary:

$$\xi = 0 \quad \text{or } r = 0 \tag{8.30b}$$

Along each contour line, p_*, ρ_* and v_* are constants that, strictly speaking, depend on γ and $\gamma = 1.4$ for air.

Investigating the flow field behind a shock wave front requires use of independent variables r and t to solve a system of partial differential equations for conservation of mass, conservation of momentum and energy. Based on dimensional analysis, flow field depends on one independent variable: dimensionless ξ. Therefore, instead of having to solve partial differential equations, it is necessary to solve only a system of ordinary differential equations. Obviously, it is much simpler and easier to judge that the flow field has self-similarity characteristics that one part of the flow field resembles another part of the field. In particular, spacial distributions of state variables including pressure, density and particle velocity at $t = t_1$ resemble spacial distributions at $t = t_2$ and variations of those stated variables with time at $r = r_1$ resemble variations at $r = r_2$.

Considering the importance of the subject and academic ingenuity, the analyses and contributions of Taylor et al. concerning intense explosion of point sources compare favorably with Prandtl's similarity solution for the theory of boundary layer and may be rated among the finest work in mechanics.

8.2 Explosive Working

Explosion is a process of energy release that is accompanied by generation of high pressure, high particle velocity and extremely high power density. Such intense dynamic loading is used in various mechanical operations, such as forming thin sheet metal parts, welding various metallic materials, hardening surfaces of materials and synthesizing or modifying material properties. All such explosive work features short duration, high pressure and high deformation rate.

8.2.1 Explosive Forming

Explosive forming allows deep drawing, bulging, fringe crimping, bending and sizing (Fig. 8.6). The explosive charge is usually set in a water medium for pressure transmission and a blank is positioned at a finite distance from the explosive. Explosion pressure transmits at high velocity through the medium and deforms the blank. To avoid obstructing the blank forming well close to the die wall due to gas blockage in the die chamber, a venting hole must be placed in the die to allow gas to escape from the die chamber.

In the period 1960–1963, a research group led by Cheng [4] found that the deformation process has two distinct acceleration stages. The reacceleration phenomenon was explained by showing that, after explosion, shock wave hits the blank and the blank starts its first acceleration stage. Due to the blank's inertia, negative pressure develops and causes extensive cavitation in the water. Compared with much higher shock wave pressure peak, it can be assumed that this cavitation is independent of strain rate and has pressure of approximately 0. When a spherical shell bulges, a cavitated region separates the water core from an outer shell that consists of the metal shell itself and a thin layer of attached water. Due to inertia, the water core moves toward the shell, ploughs through the cavitated region and eventually closes the gap between core and outer metal shell. Collision occurs and the metal shell reaccelerates. Later, the metal shell slows and the forming process is complete. Process details have complex geometry, but basic mechanical principles can be outlined.

In free-forming, there are no dies and only a ring holds the blank. Figure 8.7 shows two types of forming using dies: *deep drawing* and *bulging*. In deep drawing, a concentrated charge is usually set at the symmetrical axis at a finite distance from the blank. In bulging, a linear charge is set at the symmetrical axis. Detonation generates explosion products with high pressure and high velocity, resulting in a water shock wave that propagates outward. The shock wave hits and

Fig. 8.6 Explosive forming

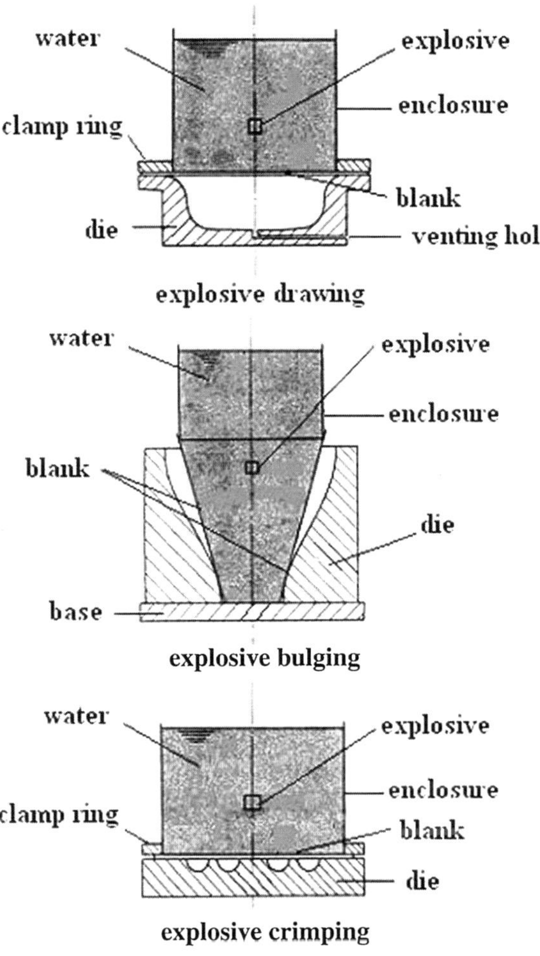

Fig. 8.7 Explosive forming with die

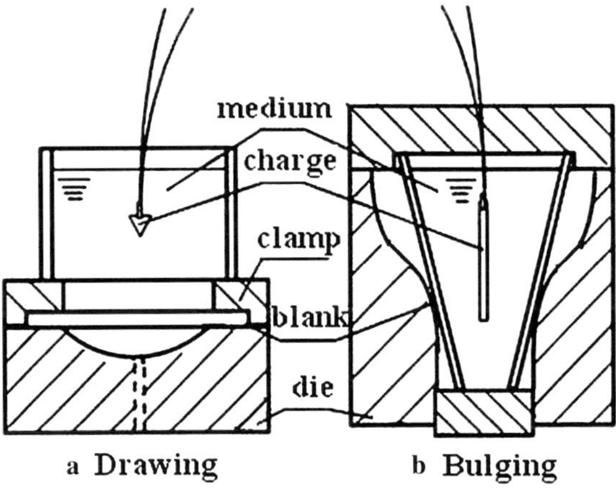

pushes the blank and that blank forms freely or forms in a die. Several factors affect governing parameters for explosive forming:

Factor 1. Explosive: mass of explosive charge Q, loading density ρ_e, chemical energy released per unit mass of explosive E_e and dilatation index of detonation products γ_e;

Factor 2. Water: initial density ρ_w, pressure coefficient B and adiabatic index γ_w;

Factor 3. Charge location: distance of the charge from the blank h;

Factor 4. Blank: density ρ, elastic constants E and v, yield limit Y, thickness δ and characteristic diameter D;

Factor 5. Die: characteristic deflection y.

1. Explosive working without dies—free-forming by explosive explosion

In free-forming, there is no die and no constraint related to die. Deformation is characterized by a function of the above governing parameters and maximum deflection y is:

$$y = f(Q,\ \rho_e,\ E_e,\ \gamma_e;\ \rho_w,\ B,\ \gamma_w;\ h;\ \rho,\ E,\ v,\ Y,\ \delta,\ D). \qquad (8.31)$$

Generally, the blank is thin and ratio of thickness to diameter δ/D is only a few percent. Inertia and deformation properties of the blank can be represented by combining parameters $\rho\delta$, $E\delta$, $Y\delta$ and v and relationship (8.31) reduces:

$$y = f(Q,\ \rho_e,\ E_e,\ \gamma_e;\ \rho_w,\ B,\ \gamma_w;\ h;\ \rho\delta,\ E\delta,\ v,\ Y\delta,\ D). \qquad (8.32)$$

Taking ρ_e, E_e and D as a unit system produces dimensionless relationship:

$$\frac{y}{D} = f\left(\frac{(Q/\rho_e)^{1/3}}{D},\ \gamma_e;\ \frac{\rho_w}{\rho_e},\ \frac{B}{\rho_e E_e},\ \gamma_w;\ \frac{h}{D};\ \frac{\rho\delta}{\rho_e D},\ \frac{E\delta}{\rho_e E_e D},\ v,\ \frac{Y\delta}{\rho_e E_e D}\right). \qquad (8.33)$$

If the explosive, material of the blank and pressure transmission medium (usually water) are the same in model and prototype, related parameters are constants:

$$(\rho_e,\ E_e,\ \gamma_e;\ \rho_w,\ B,\ \gamma_w;\ \rho,\ E,\ v,\ Y) = \text{constant}.$$

Thus, there is a simple dimensionless relationship:

$$\frac{y}{D} = f\left(\frac{(Q/\rho_e)^{1/3}}{D},\ \frac{h}{D},\ \frac{\delta}{D}\right), \qquad (8.34a)$$

where the numerator of the first dimensionless independent variable $\left(\frac{Q}{\rho_e}\right)^{1/3}$ can be regarded as characteristic length of the charge and denoted by l_e. Therefore:

$$\frac{y}{D} = f\left(\frac{l_e}{D}, \frac{h}{D}, \frac{\delta}{D}\right). \tag{8.34b}$$

Result (8.34b) shows that free-forming follows a geometric similarity law. In other words, if materials used in model and prototype are the same, it is sufficient to have a small-scale modeling test using geometry that is similar to the geometry in the prototype in order to have similar geometries in the formed parts.

2. Explosive forming with dies

If deflection in forming is considered, a procedure similar to that used in free-forming can describe forming with dies. In free-forming, charge mass is regarded as an independent variable and characteristic deflection of the work piece is regarded as being a dependent variable. In explosive forming with dies, characteristic deflection may be regarded as an independent variable and the mass of the charge may be regarded as a dependent variable. Both types of explosions allow similar discussion and drawing of conclusions about the roles of independent and dependent variables. If the same explosive, blank material and pressure transmission medium are used in model and prototype, there is dimensionless relationship:

$$\frac{(Q/\rho_e)^{1/3}}{D} = f\left(\frac{y}{D}, \frac{h}{D}, \frac{\delta}{D}\right). \tag{8.35a}$$

In forming with dies, relative characteristic deflection $\frac{y}{D}$ is given beforehand. After the work piece impacts the die, the work piece may rebound due to elastic unloading, which could result in unsatisfactory working precision. To improve this precision, a slightly larger value for $\frac{y}{D}$ may be taken, a value that relies on experience. Because $(Q/\rho_e)^{1/3}$ can be regarded as characteristic length l_e of the explosive charge, (8.35a) reduces to a *geometrical similarity law* for explosive forming with dies:

$$\frac{l_e}{D} = f\left(\frac{y}{D}, \frac{h}{D}, \frac{\delta}{D}\right). \tag{8.35b}$$

Cheng [5] considered that a thin work piece is appropriate, namely $\frac{\delta}{D} \ll 1$, making it possible to derive a *modeling law* that is simpler than the geometrical similarity law. Basically, deformation of a thin sheet work piece is a kind of membrane tension and bending energy can be ignored. In forming with dies, strain is known beforehand and the following relationship can be used to evaluate mean deformation work $\bar{w}$ based on known mean strain $\bar{\varepsilon}$:

$$\bar{w} = (Y\bar{\varepsilon}) \cdot (\delta D^2). \tag{8.36}$$

Assuming that energy efficiency is constant for a given forming geometry and explosive position relative to the work piece, the ratio of mean deformation work $\bar{w}$ to chemical energy released by the explosive charge QE_e is constant:

$$\frac{\bar{w}}{QE_e} = \frac{(Y\bar{\varepsilon}) \cdot (\delta D^2)}{QE_e} = \text{constant}. \tag{8.37a}$$

Thus, mean strain $\bar{\varepsilon}$ is:

$$\bar{\varepsilon} = \frac{\text{constant}}{(Y\delta D^2)/(QE_e)}. \tag{8.37b}$$

Because the given forming geometry has a definite relation between mean strain $\bar{\varepsilon}$ and $\frac{y}{D}$, so

$$\frac{y}{D} = f\left(\frac{Y\delta D^2}{QE_e}\right) \tag{8.38}$$

Namely, for given $\frac{y}{D}$ in the case of forming with dies:

$$\frac{Y\delta D^2}{QE_e} = \text{constant}. \tag{8.39}$$

If materials used in model and prototype are the same, Y and E_e in (8.39) remain constants and there is a formula for the mass of explosive charge:

$$Q = \text{constant} \cdot \delta D^2. \tag{8.40}$$

This *modeling law* is called as the *energy law for explosive forming*. Only a one shot modeling test based on this energy law is needed to determine the value of the constant in formula (8.40). Extensive experiments undertaken in the 1960s applied this law to free-forming and to forming with dies.

8.2.2 Explosive Welding

Explosive welding or *explosive cladding* is a class of technology used for welding metallic materials through high pressure generated by explosion. Explosive welding is commonly used to weld or clad precious metal such as stainless steel, titanium, zirconium, copper and alloys with carbon steel in order to save precious metal. By applying common technology, it is difficult to weld two metallic materials with quite different thermal properties (melting point, heat conductivity) or different hardness. However, such welding can be achieved successfully by applying explosive technology.

Figure 8.8 sketches normal explosive welding of two metallic plates. A sheet of explosive is positioned on a metal plate called a *covered plate* or *flying plate*. Another metal plate called a *base plate* is set under and parallel to the covered plate using a stand-off distance. After detonation, explosion products with high pressure drive the covered plate. The covered plate bends and accelerates abruptly with a bending angle and a normal velocity v. Then the bent part of the covered

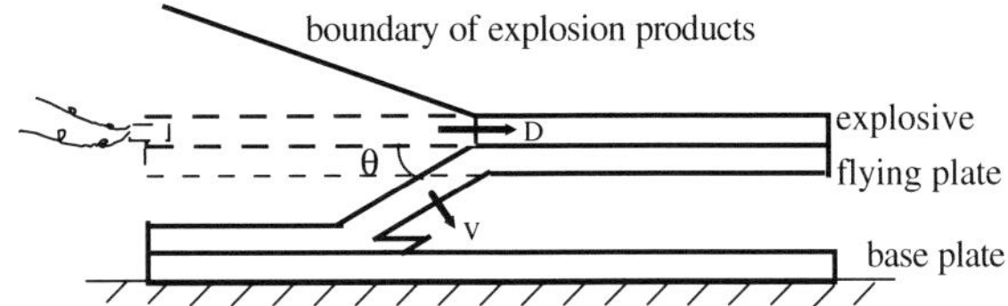

Fig. 8.8 Sketch of explosive welding

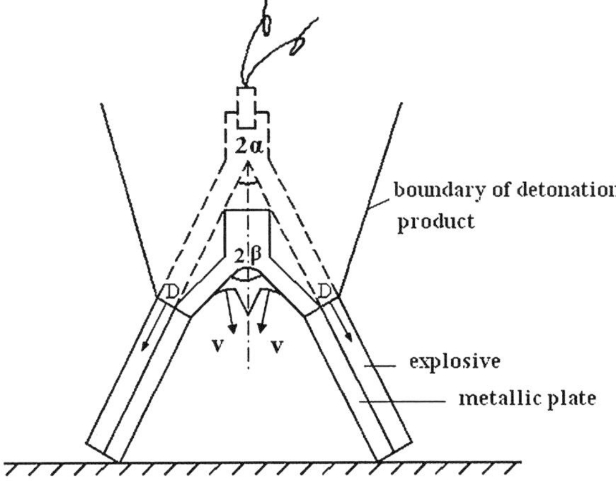

Fig. 8.9 A symmetric set-up for explosive welding

plate impacts obliquely the base plate with the velocity v ($\approx$ hundreds of meters per second). Near the collision line, pressure in the material jumps to a high value (e.g., 10 GPa), so material strength can be ignored. Part of the material in the surface layer is stripped off and discharged in the resulting planar jet. After the jet disappears, freshly active surfaces emerge to form a metallurgical combination that is subject to high pressure between covered plate and base plate.

Figure 8.9 shows a symmetrical set-up that is useful in examining explosive welding and optimal parameters for that process. Two plates of the same metallic material and same thickness δ are placed symmetrically above the ground, with the top edges of the material touching to form an arch. Initial angle between the two plates is 2α and their outer surfaces are covered with two sheets of the same explosive of the same thickness δ_e. A detonator is positioned on top of the explosive sheets.

After detonation, a detonation wave with velocity D propagates obliquely downward along both explosive sheets. Detonation products having high pressure push the metallic plates and cause the plates to bend abruptly. After this bending, the normal velocity of the plates jumps from 0 to v and the corresponding bending angle $\theta = \arcsin\dfrac{v}{D}$. The angle between the two plates increases from initial static value 2α to dynamic angle 2β, so $2\beta = 2\alpha + 2\theta$. The bent parts of the two plates collide at the symmetrical plane with a high relative velocity and compress each other. Adjacent to the collision line, some material in the thin surface layers is squeezed away and a sheet-like jet forms. After this jet disappears, fresh active surfaces emerge that form a wave interface and there is a well combination in solid phase state. The formation of new interface and the well combination induced by the formation and disappearance of the jet are keys in successful welding. Modeling tests are designed to find proper explosives and optimal governing parameters, including mass of explosive, static angle (or stand-off distance) between the metallic plates.

Experiments show that welding quality correlates with wave interface shape and well welding corresponds to a wave interface that has low amplitude and proper wave length. Collision with extremely high relative velocity produces partial melting in the surface layer and forms a wave interface that has high amplitude. The melted region becomes a shrunken hole due to cooling that negatively affects welding quality. Collision with insufficient relative velocity produces pressure that is too low to form a jet sheet, so the resultant interface is flat and easily separated.

Experiments show that a wave interface with small amplitude and proper wave length are essential in successful explosive welding. Wave formation conditions are crucial in the mechanism of such welding.

The most common arguments in the 1980s were either based on the Kármán vortex street theory [6] or on the Helmholtz theory of instability of parallel flows [7]. However, there are strong reasons for believing that conditions assumed to apply in these theories do not actually apply, especially in view of experiments [8]. Viscosity does not play an important role since geometric similarity law applies and interfacial waves form very close to the stagnation point where parallel flow theory does not apply. Most existing theories are based on a pure fluid model and are not convincing. In fluids, motion eventually subsides when overall motion stops, leaving no noticeable marks. In contrast, late stage motion of hydro-elasto-plastic media is solid-like, so certain features of the previous motion freeze, remain permanent and are observable after the event.

A fluid–solid model provides insight into interface wave formation [8]. Applying dimensional analysis to formation of the interfacial waves allows the welding process to be divided into an earlier stage when explosive detonation pushes the covered plate and a later stage when the covered plate impacts the base plate with high velocity and the plates weld.

Stage 1. When detonation products having high pressure exert force on the covered plate that covered plate bends abruptly with a bending angle at the detonation wave front θ. Normal velocity of the plate is $v = D \cdot \sin \theta$, where D = detonation velocity. Because detonation pressure is much higher than strength of the metallic material, the covered plate can be regarded as a fluid. Assuming that thickness of the explosive sheet $= H_e$, loading density $= \rho_e$, detonation velocity $= D$, the dilatation index of detonation products $= \gamma_e$ and assuming that thickness of both metallic plates $= H$, density $= \rho$, sound velocity $= c$, and dilatation index $= \gamma$, normal velocity of the driven covered plate v:

$$v = f(H_e, \rho_e, D, \gamma_e; H, \rho, c, \gamma). \tag{8.41}$$

Since $\theta = \arcsin \dfrac{v}{D}$, bending angle is also a function of the above parameters. Taking H, ρ and D as a unit system:

$$\frac{v}{D} = f\left(\frac{H_e}{H}, \frac{\rho_e}{\rho}, \frac{c}{D}, \gamma_e, \gamma\right). \tag{8.42}$$

Judging from the point of view of physics, explosive welding is a process of energy transfer in which chemical energy of the explosive sheet changes to kinematic energy of the covered plate. To express such transfer, the first two dimensionless parameters in (8.42) can combine to form a new dimensionless parameter of substance called *specific mass of charge*:

$$\frac{v}{D} = f\left(\frac{H_e \rho_e}{H \rho}, \frac{c}{D}, \gamma_e, \gamma\right). \tag{8.43}$$

For given explosive and plate material, specific mass of charge $\frac{H_e \rho_e}{H \rho}$ is a unique dimensionless parameter for determining relative velocity $\frac{v}{D}$.

Stage 2. When two flying plates collide with normal velocity v and dynamic angle 2β, deformation of those metallic plates depends on *inertia* and strength of the plates. In such a case, the metallic plates should be regarded as hydro-elasto-plastic bodies. However, because deformation is so large that elastic effect of metal can be ignored, thickness of metal plates H, density ρ and yield limit Y may be regarded as parameters for the plates. Thus, wave length of the interfacial wave:

$$\lambda = f(v, \beta; H, \rho, Y). \tag{8.44}$$

Taking H, ρ and Y as a unit system:

$$\frac{\lambda}{H} = f\left(\frac{v}{(Y/\rho)^{1/2}}, \beta\right). \tag{8.45}$$

Zhang, Li et al. [9] presented experimental results for the variation of relative wave length $\frac{\lambda}{H}$ with $\frac{\rho v^2}{2Y}$ and β in the ranges $2\beta = 12°$, $18°$ and $26°$, and $6.5 \leq \frac{v}{(Y/\rho)^{1/2}} \leq 18$ (Fig. 8.10). The abscissa in the figure $= \frac{\rho v^2}{2Y}$ and corresponds to $\frac{v}{(Y/\rho)^{1/2}}$. These results show good correlation between relative wave length $\frac{\lambda}{H}$ and relative dynamic pressure $\frac{\rho v^2}{2Y}$. Experiments verify that geometrical similarity law do hold and those experiments focus on a large range of variation related to thickness of plates, i.e., ratio of maximum and minimum thickness $= 10$. Experimental results clearly show the influence on interfacial wave phenomenon of strength that is independent of strain rate. However, those experiments do not

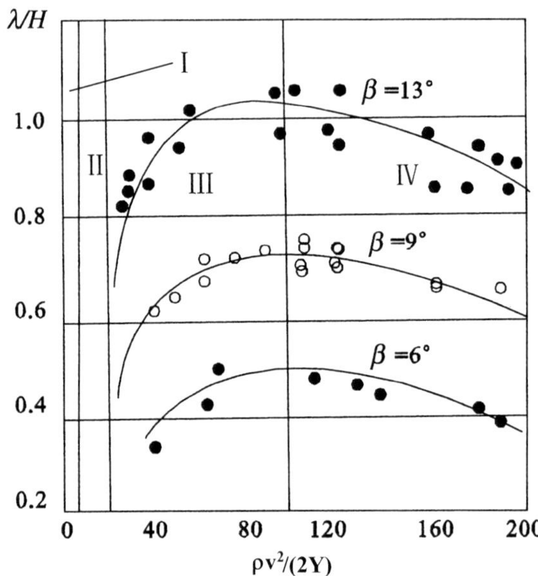

Fig. 8.10 Variation of λ/H with $\rho v^2/(2Y)$ and β

prove the validity of the model of viscous fluid. In particular, relative dynamic pressure $\frac{\rho v^2}{2Y}$ or *damage number* $\frac{\rho v^2}{Y}$ are key similarity criterion numbers.

Fig. 8.10 shows four regimes of different features:

Regime I. $\rho v^2/2Y < 10$: no welding occurs;

Regime II. $10 \le \rho v^2/2Y \le 23$: interface is flat and not wavy;

Regime III. $23 \le \rho v^2/2Y \le 95$: interface is wavy and λ/δ increases with $\rho v^2/2Y$;

Regime IV. $\rho v^2/2Y > 95$: λ/δ tends to decrease.

Synthesizing features of Regimes I–III show that, compared to strength, inertia becomes increasingly important as $\rho v^2/2Y$ increases. Cheng and Li [10] consider that relative wave length λ/δ in Regime IV depends on inertia, strength and compressibility of metallic material. In other words, an additional governing parameter is *Mach number v/c*, where c = characteristic sound velocity of the material.

8.3 Blasting

Blasting is a dynamic phenomenon that occurs when charges set in a medium of rock, soil or an engineering structure explode and cause deformation, damage, loosening, disintegrating and ejection of medium or structure.

The process of blasting roughly divides into several stages: detonation of explosive charge, propagation of explosion waves in medium, formation and

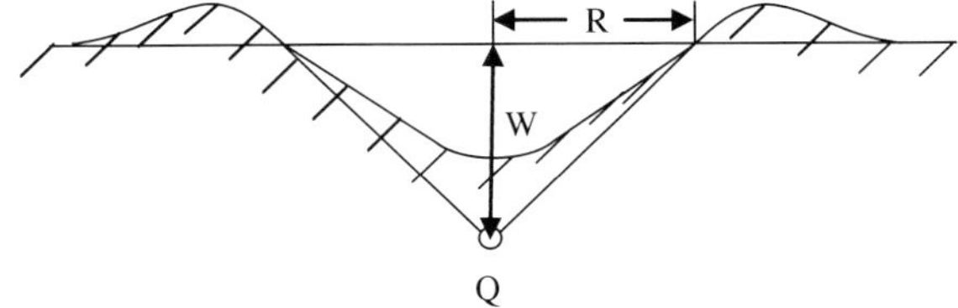

Fig. 8.11 Blasting crater for concentrated charge

expansion of a dome, ejection of parts of the medium and the falling down and piling up of those parts.

Blasting is complicated and the design and undertaking of small-scale modeling tests using dimensional analysis requires familiarity with site topography, geology and engineering requirements. To design modeling test and guide engineering operation, the most important is to work out a proper scheme for charge distribution. There are three fundamental schemes for charge distribution:

Scheme 1.   Concentrated charge
Scheme 2.   Linear charge
Scheme 3.   Planar charge

8.3.1 Boleskov (Болесков) Formula for Blasting Using the Concentrated Charge Scheme

The selected explosive and loading density must correspond to the given medium of rock or soil. Main parameters for blasting operations that employ a concentrated charge are burial depth W and mass of explosive Q. The burial depth is the shortest distance from the charge center to the earth's surface (free surface). In engineering, burial depth is commonly called "*the line of least resistance (LLR)*." After the explosive charge is initiated, energy releases, medium deforms and even damages, there is work against gravity, medium elevates and even launches into the air, eventually falling and piling. Apparently, resistance is weakest along the direction of LLR.

Figure 8.11 shows blasting in rock or soil with a concentrated charge. For the explosive charge, it is assumed that loading density $= \rho_e$, chemical energy released per unit mass of explosive $= E_e$ and dilatation index of detonation products $= \gamma_e$. For the medium, it is assumed that density $= \rho$, elastic constants $= E$ and v, strength (probably tensile strength or shear strength) of damage $= S$ and dilatation index $= \gamma$. If LLR $= W$ and mass of concentrated explosive charge $= Q$, blasting causes damage to the medium and ejects parts of the medium to form a blast crater on the earth's surface. In engineering, this crater is called *crater in sight*. The radius of the circle intersecting the crater and the free surface is defined as radius of the crater in sight, denoted as R and the ratio of R to LLR is defined as the blasting action index, denoted as n:

$$n = \frac{R}{W}. \tag{8.46}$$

Radius of the crater in sight is a function of above parameters related to charge, medium and buried depth:

$$R = f\left(Q,\ \rho_e,\ E_e,\ \gamma_e;\ \rho,\ E,\ v,\ S,\ \gamma;\ W\right). \tag{8.47}$$

Taking ρ_e, E_e and W as a unit system:

$$\frac{R}{W} = f\left(\frac{(Q/\rho_e)^{1/3}}{W},\ \gamma_e,\ \frac{\rho}{\rho_e},\ \frac{E}{\rho_e E_e},\ v,\ \frac{S}{\rho_e E_e},\ \gamma\right), \tag{8.48a}$$

or

$$n = f\left(\frac{(Q/\rho_e)^{1/3}}{W},\ \gamma_e,\ \frac{\rho}{\rho_e},\ \frac{E}{\rho_e E_e},\ v,\ \frac{S}{\rho_e E_e},\ \gamma\right), \tag{8.48b}$$

If explosive and medium adopted in model and prototype are the same, (8.48b) can be:

$$n = \frac{R}{W} = f\left(\frac{(Q/\rho_e)^{1/3}}{W}\right). \tag{8.49}$$

Because $(Q/\rho_e)^{1/3}$ can be regarded as characteristic length of the charge, expression (8.49) represents a *geometrical similarity law*.

According to an engineering convention and considering that density of explosive ρ_e adopted in model and prototype are the same, if units R and W are assigned by using m, the unit of Q kg respectively, then expression (8.49) can reduce to a relationship between magnitudes of the related quantities:

$$n = \frac{R}{W} = f\left(\frac{Q^{1/3}}{W}\right), \tag{8.50a}$$

or

$$\frac{Q}{W^3} = K \cdot f(n), \tag{8.50b}$$

where constant K represents a synthesized property of the explosive and medium used and the unit of this constant is determined by units assigned to Q and W. The particular form of function $f(n)$ should be determined by modeling tests and, based on experiments, Boleskov (Болесков) suggested that $f(n) = 0.4 + 0.6n^3$:

$$Q(kg) = K\left(kg/m^3\right) \cdot \left(0.4 + 0.6n^3\right) \cdot \left(W(m)\right)^3, \tag{8.51}$$

Analyses and experiments show that the Boleskov formula (8.51) can be applied only in cases where LLR is relatively small. Engineering practice shows that when LLR > 25 m, gravity effect should be considered and gravitational acceleration g is also a governing parameter. Adding dimensionless parameter $\frac{QE_e}{\rho_e g W^4}$ produces a dimensionless expression for the relative radius of the crater in sight:

$$n = \frac{R}{W} = f\left(\frac{Q}{\rho_e W^3},\ \frac{QE_e}{\rho_e g W^4},\ \gamma_e,\ \frac{\rho}{\rho_e},\ \frac{E}{\rho_e E_e},\ v,\ \frac{S}{\rho_e E_e},\ \gamma\right). \tag{8.52}$$

The first two independent variables in the right hand side of (8.52) relate to explosive charge Q. Variable $\frac{Q}{\rho_e W^3}$ is a ratio of two volumes but variable $\frac{Q}{\rho_e W^3} \cdot \frac{E_e}{g W}$ is not, so the geometrical similarity law is invalid if there is a large LLR.

8.3.2 Excavation Blasting for Trenches or Tunnels

Linear charges of high explosives are used in blasting operations for excavating trenches or tunnels. In such cases, it is common to use mass of explosive per unit length of linear charge q as a main parameter having dimension M/L.

In (8.3.1) Q is used to discuss blasting with concentrated charges. In discussing blasting for trenches, q of the linear charge is used instead of Q to produce an expression for the radius of trench R:

$$R = f\left(q,\ \rho_e,\ E_e,\ \gamma_e;\ \rho,\ E,\ v,\ S,\ \gamma;\ W\right). \tag{8.53}$$

Using ρ_e, E_e and W as a unit system:

$$\frac{R}{W} = f\left(\frac{(q/\rho_e)^{1/2}}{W},\ \gamma_e,\ \frac{\rho}{\rho_e},\ \frac{E}{\rho_e E_e},\ v,\ \frac{S}{\rho_e E_e},\ \gamma\right). \tag{8.54}$$

If explosive and medium adopted in model and prototype are the same, the dimensionless relationship in (8.54) reduces:

$$\frac{R}{W} = f\left(\frac{(q/\rho_e)^{1/2}}{W}\right). \tag{8.55}$$

or

$$q = \rho_e W^2 f(n),$$

where $n = \frac{R}{W}$. It should be noted that the mass of the explosive used per unit length of excavation q is proportional to W^2. In blasting using a concentrated charge, the mass of explosive used Q is proportional to W^3 but the dimensions of q and Q differ and their difference is length L.

Similar results can be obtained by using energy analysis. Assuming that energy used to deform and fracture rock or soil per unit volume $= E_s$, if LLR is relatively small and gravity effect is negligible, the ratio of the radius of trench R to LLR W, $\frac{R}{W}$, depends on ratio of explosive energy released per unit length of excavation qE_e to energy required to deform and fracture medium W^2E_s, $\frac{qE_e}{W^2E_s}$. If explosive and medium used in model and prototype are the same, $\frac{E_e}{E_s}$ is constant and there is a formula for the mass of explosive in the blasting of the linear charge:

$$q = K'W^2 \cdot f(n), \tag{8.56}$$

where $n = \frac{R}{W}$, $K' =$ dimensional characteristic quantity representing the properties of explosive and medium. The particular form of function $f(n)$ is determinable by modeling tests.

If blasting occurs in an infinite medium, such as in tunneling operations, characteristic length representing blasting effect (e.g., radius of the loosened region) is proportional to radius of the linear charge.

8.3.3 Directed Ejection Blasting of Planar Charge

For deep-hole bench blasting in open pit mining and in other blasting operations, an array of linear charges is disposed in a plane and the array is detonated simultaneously to form a planar charge group blast. In this case, supposing that energy used to deform and fracture per unit volume of the medium $= E_s$, mass of explosive distributed per unit area of charge-disposed plane q_A is normally a main parameter with dimension M/L^2. If the LLR is relatively small and gravity effect is ignored, the geometrical similarity law still holds. In directed ejection blasting, mean flight velocity $\bar{v}$ representing blasting effect depends on ratio of energy of explosive q_AE_e to energy used to deform and fracture medium per unit area of excavation surface WS, $\frac{q_AE_e}{WS}$, where S is medium strength. If explosive and medium used in model and prototype are the same, there is a formula for the mass of explosive charges used per unit distributed area:

$$q_A = K''W \cdot f(n), \tag{8.57}$$

where $n = \frac{\bar{v}}{E_e^{1/2}}$, $K'' =$ characteristic dimensional quantity representing properties of explosive and medium. The particular form of function $f(n)$ can be determined by modeling tests and mass of explosive used per unit area of excavation q_A is proportional to W.

Directed ejection blasting in soil with horizontal free surface is difficult because that type of surface is a strong planar sink. Ordinary explosive distribution causes a towering blast, meaning that sinking or suctioning play due to the horizontal free surface cannot be underestimated.

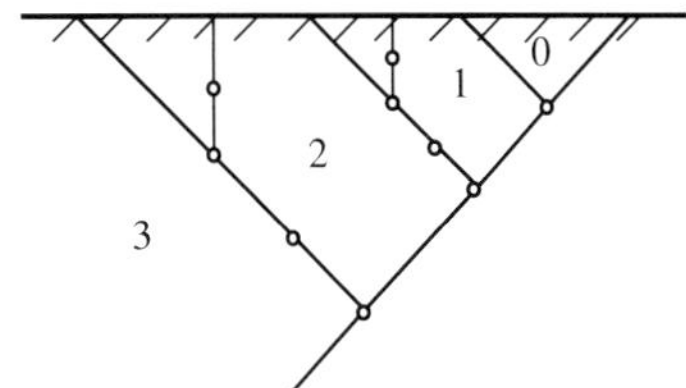

Fig. 8.12 Multiple row delayed action directed blasting

Systematic small-scale field experiments study the transport of rock or soil mass from flat horizontal ground and create a multiple-row delay blasting scheme with arrays of explosive charges [11]. The design is based on the fact that horizontal free surface terrain is unfavorable for directed ejection blasting and that a free surface with 45° angle inclination is the most favorable terrain. A *multiple-row delay blasting scheme* is used in which the blasting of a row creates favorable terrain with a 45° inclination for a succeeding row that is exploded after proper delay (Fig. 8.12). In this scheme, a trench with side walls inclined at 45° is excavated manually or through blasting in advance. On one side of the trench, a proper distance W_1 is selected as the LLR for blasting the first row. Along a plane that is parallel to but distant from this side wall of the trench, an array of linear charges is disposed with proper spacing so that blasting effect approaches that induced by a plate-like charge. In actual operations, each linear charge is replaced by an array of properly spaced concentrated charges. The free surface for the first row consists of a horizontal part on top and an inclined part on the side, with the inclined part designed to be dominant. According to the same principle, a proper LLR W_2 (m times W_1 and $m > 1$ in general) is selected for the second row. An array of charges is disposed along a plane that is parallel to the plane of the first row of arrayed charges. The initiation for neighboring rows is separated by proper timing using delayed-action detonators. Subsequent rows are arranged according to the same principle.

In small-scale modeling tests, several simple and easy but effective techniques can measure displacement and velocity were developed. In this testing, small specific chunks (made from concrete) called *markers* are numbered and buried in soil at prescribed positions. Parts of concrete chunks inserted with pyrotecnic tablets. In the ejection process after ignition of charges, color orbits of the pyrotecnic tablets can be photographed and the velocity of flying soil fragments can be obtained. After blasting, numbered chunks that fall back to earth can be recovered and used to record flight distance. Such marker data can be used to draw a diagram of contours of flight equidistance. Comparing this diagram with a diagram of ideal contours of equidistance of flight can be used to optimize the charge distribution scheme.

The following section describes and analyzes blasting of a single row with planar-disposed charges, blasting of multiple rows of delayed-action charges and relation between ejection distance and specific explosive consumption.

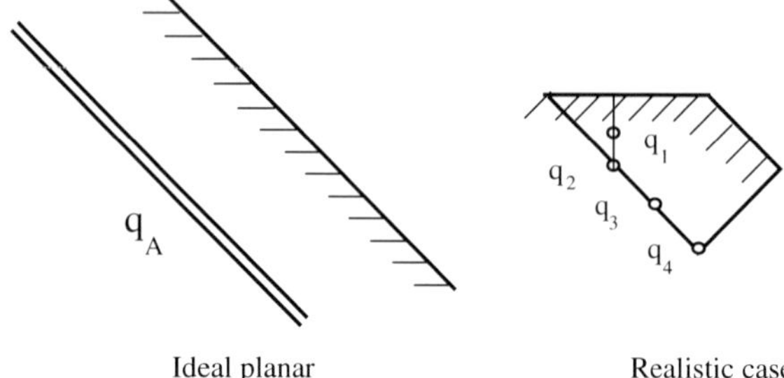

Fig. 8.13 A sketch of comparison between an ideal planar case (left) and a realistic case (right)

1. Blasting of a single row with planar-disposed charges

If the free surface is inclined at an angle of 45°, a plate-like explosive charge may be disposed in a plane parallel to the declined free surface with W of the LLR. Sufficient specific mass of explosive q_A (per unit area of the surface upon which charges are disposed) can be used so that the prescribed soil mass that is positioned obliquely above that plane has perfect flight velocity and flight distance. In an ideal case of free surface, the contours showing equidistant flight must be a set of planes that are parallel to the free surface.

In blasting of a single row, an array of charges is disposed on one side of the prescribed trench in a plane at an inclination angle of 45° and charges are ignited simultaneously. Blasting can be expected to produce perfect ejection and favorable terrain for blasting of the next row, if necessary. Figure 8.13 shows an ideal case of blasting using a single plate-like charge q_A with an infinite free surface (left) and a realistic case of blasting using a set of linear charges q_1–q_4 with both horizontal and inclined parts of free surface (right).

In the plane upon which charges are disposed, four linear charges with proper spacing q_1, q_2, q_3 and q_4 are disposed as a substitute for a plate-like charge. To obtain optimal flight distance, distribution of the explosive mass between q_1 and q_4 is determined by experiments. In the operations, each linear charge is replaced by an array of concentrated charges along the horizontal direction.

Experiments show that the directional effect of the top charge q_1 is weak because its distance from the horizontal part of the free surface is shortest and the portion of soil mass that is over q_1 is ejected almost vertically. Therefore, it is more effective to dispose q_1 forward and to the right at the top of q_2 in order to strengthen the directional effect for soil mass between the $q_1 q_2$ plane and the free surface.

Figure 8.14 shows charge distribution for a blast in saturated clay and contour lines of flight equidistance calculated from marker data. This figure shows that contours of soil mass in the prescribed region are basically parallel to the inclined free surface. Therefore, blasting approaches ideal effectiveness using an infinite free surface that is inclined at 45°.

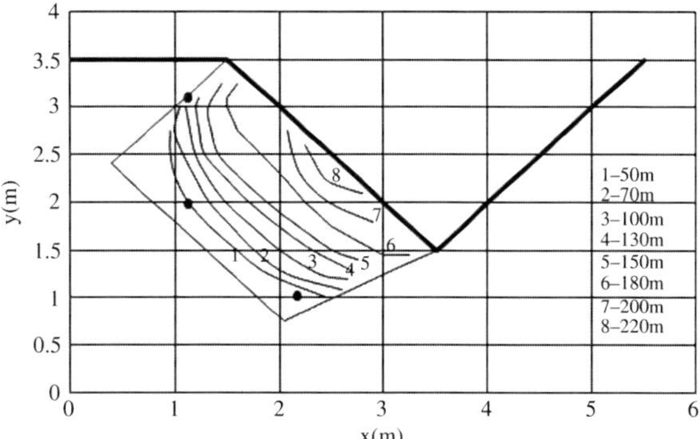

Fig. 8.14 Distribution of contour lines of flight equidistance

2. Blasting of multiple rows using delayed-action charges

Charge distribution adopted for succeeding rows are geometrically similar to distribution used for the first row. The delay time of detonation τ between adjacent rows should be properly selected. A succeeding row should be detonated after the soil mass of the preceding row has lifted and separated completely from the earth's surface and before the soil mass of the preceding row falls back to the earth's surface. Delay time can be selected in the range:

$$\frac{W}{v} < \tau < \frac{v}{g}, \tag{8.58}$$

where $W = \mathrm{LLR}$, $v = $ mean flight velocity of the preceding row and $g = $ gravitational acceleration.

3. Relation of ejection distance and specific explosive consumption

Directional blasting is complicated and involves detonation of explosives, deformation, fragmentation, acceleration and effects of gravity and air resistance in the ejection process. Applying dimensional analysis to the whole process usually cannot provide particular and useful results.

The process of blasting and ejection can be divided into two stages.

Stage 1. Explosion and dome rupture: starts at detonation of explosive charges, continues until the rupture of the dome and ends with the formation of a group of soil fragments that have a certain distribution of flight velocities and sizes.

Stage 2. Ejection and piling: the group of soil fragments with initial velocity distribution is thrown into the air and affected by gravity and air resistance until fragments fall to earth's surface and pile up in a mound of certain shape.

Contour lines of flight equidistance are basically parallel to the inclined part of the free surface, so it may be assumed that the motion is approximately one-dimensional.

In the first stage, the blasting effect is mainly to make soil fracture and form fragments and resultant ejection velocity is independent of gravity and air resistance. Gravity and air resistance are irrelevant to soil fragmentation and acceleration when LLR < 25 m. Governing parameters relate to three factors:

Factor 1. Explosive: mass of explosive per unit area of disposal surface q_A, loading density ρ_e, chemical energy released per unit mass of explosive E_e and dilatation index of detonation products γ_e;

Factor 2. Soil: density ρ, elastic constants E and v and strength S;

Factor 3. Explosive disposal surface: LLW W and inclination angle α.

Therefore, mean flight velocity of soil fragments $\bar{v}$ depends on the above parameters:

$$\bar{v} = f(q_A,\ \rho_e,\ E_e,\ \gamma_e;\ \rho,\ E,\ v,\ S;\ W,\ \alpha). \tag{8.59}$$

Taking ρ_e, E_e and W as a unit system produces dimensionless relationship:

$$\frac{\bar{v}}{E_e^{1/2}} = f(\frac{q_A}{\rho_e W},\ \gamma_e,\ \frac{\rho}{\rho_e},\ \frac{E}{\rho_e E_e},\ v,\ \frac{S}{\rho_e E_e},\ \alpha). \tag{8.60}$$

If explosive and soil adopted in model and prototype are the same, the above relationship reduces:

$$\frac{\bar{v}}{E_e^{1/2}} = f\left(\frac{q_A}{\rho_e W},\ \alpha\right). \tag{8.61}$$

If unit of $\bar{v}$ is assigned to m/s and unit of $\frac{q_A}{W}$ is assigned to kg/m^3, then (8.61) reduces:

$$\bar{v} = f\left(\frac{q_A}{W},\ \alpha\right). \tag{8.62}$$

For the case $\alpha = 45°$, mean flight velocity depends only on specific explosive consumption $\frac{q_A}{W}$, which is the mass of explosive consumed per unit volume of ejected soil fragments.

In the second stage, ejected soil fragments of a certain velocity distribution fly into the air, fall down and pile up on the earth's surface under gravity. Experiments show that for directional blasting in clay, most ejection fragments are >10 cm in size and the order of magnitude of their flight velocities is 10 m/s. In the range of the corresponding Reynolds number, air resistance effect can be regarded as negligible. Therefore, normal ejection formula applies and ejection distance of the mass center of the ejected body:

$$L = \frac{\bar{v}^2}{g} \cdot \sin 2\alpha. \tag{8.63}$$

In the case of $\alpha = 45°$, ejection distance is the maximum:

$$L = \frac{\bar{v}^2}{g}. \tag{8.64}$$

Synthesizing results obtained in above analyses for these two stages produces:

$$\bar{v} = f\left(\frac{q_A}{W}, \alpha\right), \quad \text{and } L = \frac{\bar{v}^2}{g} \cdot \sin 2\alpha. \tag{8.65}$$

Combining these equations produces:

$$L = \frac{1}{g} \cdot f\left(\frac{q_A}{W}, \alpha\right). \tag{8.66}$$

Thus, mean ejection distance depends on mass of explosives consumed per unit volume of ejected soil $\frac{q_A}{W}$ and inclination angle favorable for ejection α.

Selecting $\alpha = 45°$ makes it possible to obtain maximum distance:

$$L_m = \frac{1}{g} \cdot f\left(\frac{q_A}{W}\right). \tag{8.67}$$

Field experiments for blasting a single row of ammonia dynamite in saturated clay using LLR $W = 1.35$ m show that ejection distance L_m increases with *specific explosive consumption* $\frac{q_A}{W}$. In the range of $\frac{q_A}{W} < 4$ kg/m^3, L_m is approximately proportional to $\frac{q_A}{W}$. When $\frac{q_A}{W} > 4$ kg/m^3, the relation curve gradually deviates from the straight line and becomes curved. Energy efficiency also decreases gradually.

References

1. Taylor, G.I.: The formation of a blast wave by a very intense explosion. Tech. Rep. RC-210, Civil Defense Research Committee, 27, June, (1941)
2. Von Neumann, J.: The point source solution. Tech. Rep. AM-9, National Defense Research Council, Div. B, 30, June, (1941)
3. Sedov, L.I.: Propagation of Intense shock waves. Appl. Math. Mech.**10**(2), 241–250 (1946) (in Russian: Седов Л И. Распространение сильной ударной волны. ПММ.**10**(2), 241–250 (1946))
4. Cheng, C.M., et al.: Explosive Working. Defense Industry Publ. Co., Beijing (1981) (in chinese)
5. Cheng, C.M., et al.: The bulging deformation and the energy criterion of a spherical shell. Tech. Rept. 6451, Chinese Scientific and Technological Committee, (1964) (in chinese)
6. Cowan, G.R., Bergman, O.R., et al.: Mechanism of bond zone wave formation in explosive-clad metals. Mat. Trans. **2**, 3144–3155 (1971)

7. Utkin, A.V., Dremin, A.N., Mihailov, A.N., Gordopolov, U.A.: Wave formation in high velocity impact of metallic plates. Phys. Combust. Explos.16(4), (1980) (in Russian: Уткин, А В, Дремин А Н, Михайлов А Н, Гордополов Ю А. Волнообразование при Высокоскоростом соударений Металлов. Физика Горения и Взрывы, 16, Вып. 4, (1980))
8. Cheng, C.M., Tan, Q.M.: Mechanism of wave formation at the interface in explosive welding. UCSD Regents' Lectures. (1984). Collected Works of Cheng C.M., pp. 415–429, Science Press, Beijing (2004)
9. Zhang, D.X., Li, G.H., Zhou, Z.H., Shao, B.H.: Effects of material strength on the interfacial wave formation in explosive welding. Acta. Mech. Sin. 16(1), 73–80 (1984)
10. Cheng, C.M., Li, G.H.: Effects of strength and compressibility of materials on wave formation in explosive welding. In: Zheng Zhemin, Cheng, C.H., Ding Jing (eds.) Proceedings of the Symposium on Intense Dynamic Loading and Its Effects. Science Press, Beijing, China. pp. 854–859 (1986)
11. Tan, Q.M., et al.: Orientational throwing blasting for a horizontal ground surface. Res. Rept., Institute of Mechanics, CAS. (1967). (in Chinese)

Chapter 9
Similarity Laws for High Velocity Impacts

This chapter introduces dynamic damage to armor produced by three types of *antitank projectiles* (*rod projectiles, shaped charges* and *projectiles for armor-spalling*) and by *hypervelocity impact, tensile fracture* of shaped charge jets, tensile fracture of thin sheets and *coal and gas outbursts*. Cases related to high velocity impacts are introduced on the basis of the hydro-elasto-plastic model. Further, there is discussion of tensile fracture of a metallic jet with high velocity, thin plate fracture and tensile damage in single phase materials. Reference is made to tensile damage of coal containing gases, a type of two-phase media that is subject to the action of tensile waves and that can develop into tensile fracture and even outburst.

Impact between two bodies with high relative velocity is a phenomenon of energy transformation from mechanical energy (kinetic energy) into another type of mechanical energy (e.g., potential energy of deformation) and thermal energy. Impact is accompanied by shock waves, high pressure, high temperature and high deformation rate that results in serious deformation, damage or even melting and vaporization of impacted bodics. Meteorite impact with Earth, impact of meteorites or space junk with spacecraft and penetration of projectiles or shaped charge jets into armor are all high velocity impact phenomena. In another type of impact, potential energy or kinetic energy transforms and causes deformation or fragmentation energy. This impact type produces abrupt tensile deformation and fracture, for example, tensile fracture of jets formed by shaped charges, bulging fragmentation of blasting cartridges and bursts of coal and gas.

9.1 Rod Projectiles

The antitank rod projectile came into service in the 1960s. Compared with conventional projectile length–diameter ratio of about 4, the rod projectile has length–diameter ratio of about 15 and projectile velocity of about 1.7 km/s, nearly twice

Q.-M. Tan, *Dimensional Analysis*, DOI: 10.1007/978-3-642-19234-0_9,
© Springer-Verlag Berlin Heidelberg 2011

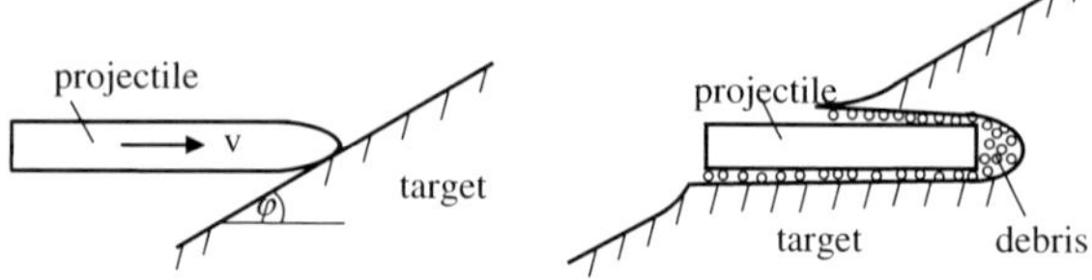

Fig. 9.1 Initial state (*left*) and intermediate state (*right*) target

the velocity of about 0.8 km/s of conventional projectiles. Rod projectile penetration depth exceeds that of conventional projectiles by as much as 1.7 times.

Figure 9.1 shows initial and intermediate states for a long rod projectile impact on a target. There is abrupt deceleration when projectile head hits target surface and a high pressure zone occurs in the region adjacent to impact point. As a result, shock waves propagate into target and projectile. Simultaneously, a crater in the target starts to form around the impact point. Unloading occurs when shock wave front reaches target surface or projectile surface. A set of rarefaction waves reflects that intercepts and causes decay in the shock wave front. In the early stage of impact, the diameter of the region that is strongly compressed by shock waves is only a few times larger than projectile diameter. Subsequently, distributions of state variables such as pressure and velocity become continuous. The crater evolves into a narrow deep hole and high pressure causes projectile head material to deform severely and fragment. Debris fragments remain close to the wall of the hole and the penetration process ends when the projectile's kinetic energy is exhausted.

It is assumed that projectile diameter $= d_p$, projectile length $= L_p$, initial velocity of projectile $= v_p$ and target inclination angle $= \phi$. It is also assumed that inertia and strength of material in projectile and target can be represented by density, elastic constants and yield limit, ρ_p, E_p, v_p, Y_p and ρ_t, E_t, v_t, Y_t, where subscript p denotes projectile and t denotes target. Because ratio of projectile velocity to sound velocity of material is relatively low, compressibility of material can be ignored. In calculating target penetration, thickness of the target being thoroughly penetrated L_t is a function of all related geometry parameters and material parameters:

$$L_t = f(d_p,\ L_p,\ v_p,\ \phi;\ \rho_p,\ E_p,\ v_p,\ Y_p;\ \rho_t,\ E_t,\ v_t,\ Y_t). \tag{9.1}$$

Taking l_p, ρ_p, and Y_t as a unit system produces dimensionless relationship:

$$\frac{L_t}{L_p} = f\left(\frac{d_p}{L_p},\ \frac{v_p}{\sqrt{Y_t/\rho_p}},\ \phi;\ \frac{E_p}{Y_t},\ v_p,\ \frac{Y_p}{Y_t},\ \frac{\rho_t}{\rho_p},\ \frac{E_t}{Y_t},\ v_t\right). \tag{9.2}$$

If materials of projectile and target in model and prototype are the same, (9.2) reduces:

$$\frac{L_t}{L_p} = f\left(\frac{d_p}{L_p}, \frac{v_p}{\sqrt{Y_t/\rho_p}}, \phi\right). \tag{9.3}$$

Apart from geometrical similarity parameters $\frac{d_p}{L_p}$ and ϕ, (9.3) contains a unique physical parameter $\frac{v_p}{\sqrt{Y_t/\rho_p}}$, the square of which, $\frac{\rho_p v_p^2}{Y_t}$, is the *damage number* that represents ratio of inertia to strength. This damage number is the most important dimensionless parameter in the hydro-elasto-plastic model. If velocities of projectiles used in model and prototype are the same, the *geometrical similarity law* holds in this case.

Gao [1] and Sun et al. [2] carried out modeling tests using reduction ratio of length $r_l = 100:19$. Comparing test results in model and prototype, the geometrical similarity law works very well and relative error for $\frac{L_t}{L_p}$ is only 2%. Since the 1970s, modeling tests based on this geometrical similarity law have become common in numerous laboratories in China.

9.2 Formation of High Velocity Jet and Jet Penetration into Target

The main part of an armor-piercing projectile is *a shaped charge*. To concentrate in a desired direction energy released by the explosive charge, a cone-shaped crater is made at one end of the charge and a thin metallic liner is inserted and placed in contact with the explosive. Due to good extensibility, red copper is often used for the liner. When a detonator initiates at the opposite end of the charge, a detonation wave propagates toward the bottom of the shaped liner. After the detonation wave reaches the top of the cone-shaped liner, high pressure detonation products squeeze liner material to form a high velocity metallic *jet* that moves in the direction of the symmetrical axis. *Jet formation* is complete when detonation wave reaches the bottom of the charge.

Figure 9.2 shows initial state of a projectile hitting the target (top), structure of a shaped charge and formation of a jet and the relatively thick tail of the jet called the *slug* (bottom). During jet formation, a ring element of liner material is pushed and compressed by high pressure of the gaseous products of detonation and accelerated in a unit time. The thickness of the ring element is constant D that equals velocity of detonation wave, but the radius of the ring element increases from minimum at the top of the cone-shaped liner to maximum at liner bottom. Correspondingly, mass of the ring element pushed by the front of the detonation wave in a unit time increases with time. Chemical energy provided by the explosive ring element that surrounds the liner element decreases due to decrease in explosive element volume. Therefore, the velocity of jet elements transformed

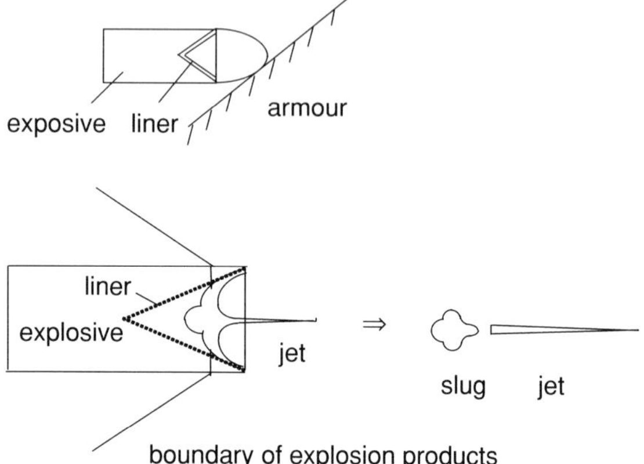

Fig. 9.2 Initial state of projectile hitting target (*top*) and jet formation (*bottom*)

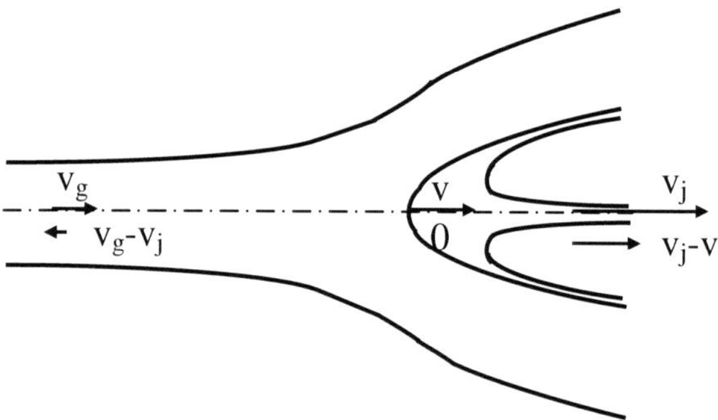

Fig. 9.3 Formation of jet with infinite length

from ring elements gradually decreases from the head to the tail of the jet. Velocity of the jet head is about 8 km/s, velocity of the tail is <1 km/s and the jet's velocity distribution is basically linear. The slug eventually separates from the main part of the long slender jet and penetration ability ends.

Taylor [3] offers a theoretical analysis of jet formation and target penetration in an ideal case in which explosive, liner and target are infinite in length. It is assumed that ultimate velocity of the jet formed by detonation $= v_j$, velocity at the intersection of symmetrical axis and interface between liner material and detonation products $= v$ and velocity of detonation products upstream $= v_g$.

Figure 9.3 shows formation of a jet having infinite length, where velocities v_g, v and v_j over the symmetrical axis are characteristic particle velocities on the symmetrical axis of coordinates fixed on the Earth. Motion is stationary in coordinates moving with the intersection point. In this moving coordinates, characteristic particle velocities $v_g - v$, 0 and $v_j - v$ are marked sequentially below the symmetrical axis.

Based on Taylor's studies, Fig. 9.4 shows penetration of a steel target with infinite thickness by a jet with velocity v_j and infinite length. The geometry is essentially the same as in jet formation. Marked over the symmetrical axis in the figure are characteristic particle velocities 0, v, v_j on the symmetrical axis that

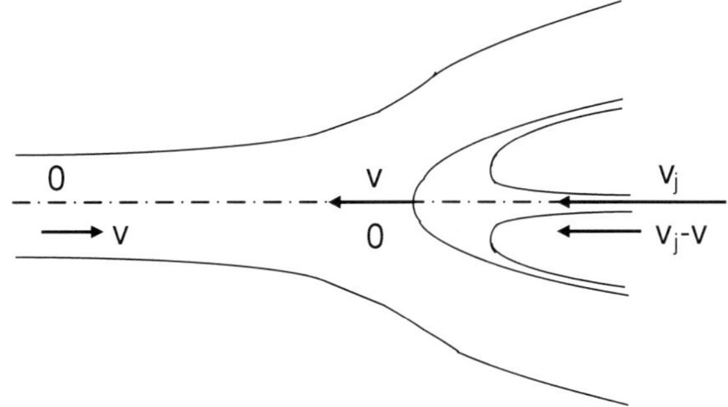

Fig. 9.4 Penetration of jet into target both with infinite length

are, respectively, velocity of the target downstream, particle velocity at the intersection point and velocity of the jet upstream. Marked below the symmetrical axis in the figure are three velocities in the coordinates moving with intersection point $-v$, 0 and $v_j - v$. Motion is stationary in moving coordinates.

Assuming that materials of jet and target are incompressible, the Bernoulli theorem fundamental in fluid mechanics can be applied. Total pressure has the same value at all points on a streamline in steady incompressible fluid flow and this total pressure is the sum of static pressure p and dynamic pressure $\frac{\rho v^2}{2}$:

$$p + \frac{\rho v^2}{2} = \text{constant}. \tag{9.4}$$

Applying (9.4) to central streamlines for both sides of target and jet:

$$p_a + \frac{\rho_t v^2}{2} = p_0 = p_a + \frac{\rho_j (v_j - v)^2}{2}. \tag{9.5}$$

In (9.5), the sum of the terms in the left hand side represents total pressure of the target downstream and static pressure = atmospheric pressure p_a. In the middle of the equation, p_0 = stationary pressure at intersection point, where velocity $- 0$. The sum of the two terms in the right hand side represents total pressure of the jet upstream and static pressure = atmospheric pressure p_a. Moving velocity of the hole bottom, *penetration velocity v*, can be derived from (9.5) to satisfy:

$$\frac{v}{v_j} = \frac{1}{1 + (\rho_t/\rho_j)^{1/2}}. \tag{9.6}$$

Because length of jet consumed in a unit time is $v_j - v$ and depth of penetration hole increase in unit time = v, ratio of penetration depth P to length of jet consumed in penetration L:

$$\frac{P}{L} = \frac{v}{v_j - v} = \left(\frac{\rho_j}{\rho_t}\right)^{1/2}. \tag{9.7}$$

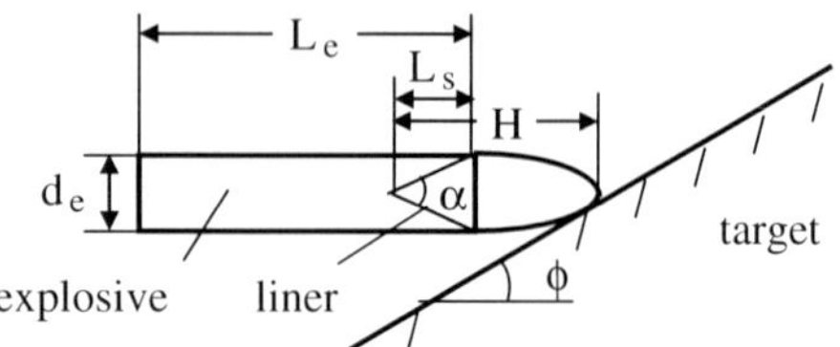

Fig. 9.5 Geometrical sketch of projectile hitting target

Steel target density is $\rho_t = 7.8$ g/cm^3. Considering that the jet undergoes extreme distortion and resultant temperature is higher than that of the original liner, density of the red copper jet is approximately $\rho_j = 8.6$ g/cm^3. Therefore, ratio of *penetration velocity* v to jet velocity v_j and ratio of penetration depth P to jet length consumed L:

$$\frac{v}{v_j} = \frac{1}{1 + (7.8/8.6)^{1/2}} = \frac{1}{1 + 0.95} = 0.51 \approx \frac{1}{2}. \tag{9.8a}$$

$$\frac{P}{L} = \left(\frac{8.6}{7.8}\right)^{1/2} \approx 1. \tag{9.8b}$$

Moving velocity of the hole bottom is roughly half of jet velocity and penetration depth in steel target P roughly equals jet length L.

Taylor analyzes the formation of a jet in a way similar to analyzing jet penetration and presents a concise and important relationship by applying the Bernoulli Theorem to formation and penetration of a metallic jet having high velocity. Relationships (9.6 and 9.7) provide theoretical basis for design of armor and armor-piercing projectiles during the Second World War and for decades after.

By the 1970s, armor strength had doubled since the Second World War and earlier models were considered to be ineffective, especially in middle and late stages of penetration when velocity of jet elements reduces linearly with time [4]. In middle and late stages, target material strength is significant and the hydro-elasto-plastic model should apply in analysis of the whole penetration process. Because penetration velocity is insufficient, effect of target material compressibility can be ignored and properties of inertia and strength are keys. Governing parameters for jet penetration into the target have three factors (Fig. 9.5):

Factor 1. Shaped charge consisting of explosive and liner.

Explosive: characteristic length L_e, loading density ρ_e, chemical energy released or unit mass of explosive E_e and dilatation index of detonation products γ_e;

Liner: characteristic length L_s, thickness δ, density ρ_s and cone angle α;

Factor 2. Target: density ρ_t, elastic constants E_t, v_t and yield limit Y_t;

Factor 3. Relative position between charge and target: standoff H and inclined angle φ.

The target used for modeling tests is thicker than estimated penetration depth, so influence of target thickness is negligible. Penetration depth P is a function of related geometry parameters and material parameters:

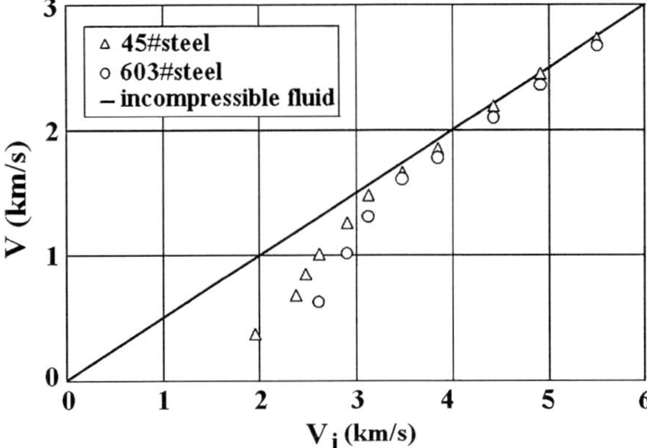

Fig. 9.6 Experimental results for v versus v_j

$$P = f(H, \ \phi; \ L_e, \ \rho_e, \ E_e, \ \gamma_e; \ L_s, \ \delta, \ \rho_s, \ \alpha; \ \rho_t, \ E_t, v_t, \ Y_t). \tag{9.9}$$

Taking L_e, ρ_t and Y_t as a unit system produces dimensionless relationship:

$$\frac{P}{L_e} = f\left(\frac{H}{L_e}, \ \phi; \ \frac{\rho_e}{\rho_t}, \ \frac{E_e}{Y_t/\rho_t}, \ \gamma_e; \ \frac{L_s}{L_e}, \ \frac{\delta}{L_e}, \ \frac{\rho_s}{\rho_t}, \ \alpha; \ \frac{E_t}{Y_t}, v_t\right). \tag{9.10}$$

If explosive and materials of liner and target used in model and prototype are the same, dimensional relationship (9.10) reduces to dimensionless relationship:

$$\frac{P}{L_e} = f\left(\frac{H}{L_e}, \ \phi, \ \frac{L_s}{L_e}, \ \frac{\delta}{L_e}, \ \alpha\right). \tag{9.11}$$

Variables in (9.11) are ratios of two lengths. Penetration depth of the metallic jet against the target follows a geometrical similarity law.

Gao et al. (1977) [5] did modeling tests of perpendicular penetration into thick steel targets using armor-piercing projectiles of three calibers: 80 mm, 100 mm and 160 mm. Test results show that penetration depths follow the geometrical similarity law and relative error <1%.

Cheng et al. (1977) [6] provide experimental results for the relation between *penetration velocity* v and jet velocity v_j, where a piercing projectile with shaped charge penetrates perpendicularly steel targets made of #45 steel ($Y_t = 0.6$ GPa) and #603 steel ($Y_t = 1.0$ GPa). In (Fig. 9.6), triangles represent #45 steel and circles represent #603 steel. The figure also shows Taylor's theoretical results based on using an incompressible model with a straight line through the origin:

$$\frac{v}{v_j} = \frac{1}{1 + (\rho_t/\rho_j)^{1/2}}.$$

Compared with experimental points, Fig. 9.6 shows that when $v_j < 4$ km/s, corresponding to $\frac{\rho_j v_j^2}{Y_t} < 10.0$, relative error for theoretical prediction based on

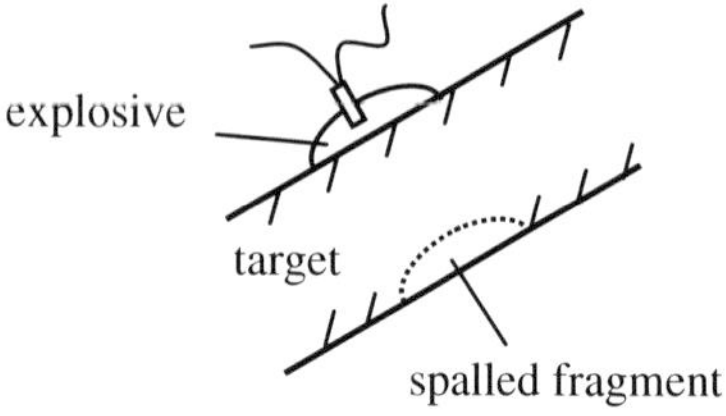

Fig. 9.7 Spallation of armor

Taylor's analysis >5%. In other words, in middle and late stages of the penetration process, when velocity of jet elements <4 km/s, effect of target strength on penetration depth must be considered.

9.3 Spallation of Armor Induced by Explosions

The main part of an armor-spalling projectile is a plastic explosive charge. When a projectile hits armor, plastic explosive deforms and extends over a certain armor surface area due to inertia. A delayed action detonator initiates plastic explosive (Fig. 9.7) and explosion produces a shock wave that propagates into the target. When shock wave front reaches rear surface of the target, a set of rarefaction waves reflects. Because the intensity of the rear of the shock wave is weaker than the intensity of the head of the rarefaction waves, tensile fractures occur in a plane parallel and adjacent to the back surface of the target. Spallation finally occurs due to the accumulation of tensile fractures. Generally, a disk-like fragment with a mass of several kilograms separates from the target with velocity that is several hundred meters per second.

If tensile strength controls spallation, a one-dimensional model can be used to analyze armor spallation. Governing parameters have two factors:

1. Explosive: thickness h_e, loading density ρ_e, chemical energy released by unit mass of explosive E_e and dilatation index of detonation products γ_e;
2. Metallic target: thickness h, density ρ, elastic constants E, v, yield limit Y and tensile strength S.

Hence, spallation thickness δ is a function of related explosive parameters and target parameters:

$$\delta = f(h_e,\ \rho_e,\ E_e,\ \gamma_e;\ h,\ \rho,\ E,\ v,\ Y,\ S). \tag{9.12}$$

Taking ρ_e, E_e, and h as a unit system:

$$\frac{\delta}{h} = f\left(\frac{h_e}{h},\ \gamma_e;\ \frac{\rho}{\rho_e},\ \frac{E}{\rho_e E_e},\ v,\ \frac{Y}{\rho_e E_e},\ \frac{S}{\rho_e E_e}\right). \tag{9.13}$$

If explosives and target materials adopted in model and prototype are the same, (9.13) reduces:

$$\frac{\delta}{h} = f\left(\frac{h_e}{h}\right). \tag{9.14}$$

It seems that spallation thickness follows a geometrical similarity law.

In the 1970s, researchers noted that spallation thickness does not follow a geometrical similarity law [7]. Experiments showed that spallation occurs not only when tensile stress exceeds tensile strength but also when spallation needs a finite duration in which micro-cracks form, grow and link to form macro-cracks and fractures. Thus, it became apparent that spallation is more of a development process than instantaneous phenomenon.

Empirical criteria of various modes of impulse-induced fracture suggests that, in addition to the parameters involved in the tensile strength criterion, an additional parameter in impulse-induced spallation criterion is characteristic damage time t_* or dimensionless parameter $\frac{h/t_*}{E_e^{1/2}}$. If explosives and target materials adopted in model and prototype are the same:

$$\frac{\delta}{h} = f\left(\frac{h_e}{h}, \frac{h/t_*}{E_e^{1/2}}\right). \tag{9.15}$$

Because t_* and E_e depend on material properties that remain constant in model and prototype, geometrical similarity law does not hold.

Wang and Bai et al. [8] analyzed spallation by using a damage mechanics model that includes nucleation and growth of micro-cracks until macro-fracture and spallation occur. These researchers claim that nucleation time is longer than growth time, so nucleation time dominates the spallation process. The researchers conclude that impulsive loading causes a certain amount of micro-damage in a finite duration of dynamic loading and this micro damage consists of nucleation, growth, full coalescence and extension of micro-fractures, leading to macro-fracture and spallation.

9.4 Hypervelocity Impacts

Impact velocity of meteorites that strike planets is generally >10 km/s, similar to impact velocity of kinetic energy projectiles that strike missiles in the *outer space*. Impact velocity of space junk striking a spacecraft is slightly lower, about 6–7 km/s. The main peculiarity of these impacts is that velocity is so high in the earlier stage of impact that impulsive pressure produced in material near the impact point is much higher than material strength, meaning that melting and vaporization may occur. These hypervelocity impacts differ from high velocity impacts.

For simulation tests in the laboratory, a two-stage light gas gun launches a projectile with high velocity. Maximum projectile velocity offered by this gun is about 8 km/s and projectile diameter is usually <1 cm. Impact-induced melting can be simulated for some materials with lower melting point such as aluminium-alloy, but it is impossible to simulate vaporization caused by impact. An electromagnetic gun launches a projectile with a higher velocity of about 13 km/s, but projectile diameter is smaller, <1 mm.

In the 1980s, data collected from much simulation testing using a two-stage light gas gun shows that a semi-spherical crater appears at the target surface when a spherical bullet hits a thick metallic target in typical hypervelocity impact. Two classes of empirical formulas were suggested in calculating crater diameter and it is commonly agreed that density (representing inertia) is the major parameter. Opinions diverge about whether the second most important parameter is material compressibility or material strength [9]. Empirical expression for relative crater depth:

$$\frac{P}{d_p} = c \cdot \left(\frac{\rho_p}{\rho_t}\right)^m \cdot \left(\frac{v_p}{v_*}\right)^n, \tag{9.16}$$

where P = crater depth, d_p = bullet diameter, ρ_p = bullet material density, ρ_t = target material density, v_p = bullet velocity. In (9.16), v_* may be considered to be material sound velocity c_t, representing compressibility effect, or considered to be relative material strength $\left(\frac{Y_t}{\rho_t}\right)^{1/2}$, representing *strength effect*, where Y_t = target material yield strength. Material compressibility can also be predicted to be important in the region near the impact point and in the earlier stage of impact, where a strong shock wave is an important factor. In the region far from impact point or in the later stage, compressibility weakens considerably and material strength dominates the event.

To determine which effect is most important Xiang (1990) carried out simulations using a two-stage light gas gun. Aluminium-alloy, copper and steel were materials used in bullets and targets and impact velocity ranges were 0.5–7.0 km/s. Empirical relationship was obtained:

$$\frac{P}{d_p} = 0.37 \cdot \left(\frac{v_p}{(Y_t/\rho_t)^{1/2}}\right)^{0.56} \cdot \left(\frac{v_p}{c_t}\right)^{0.11}. \tag{9.17}$$

Judging from values of power of dimensionless velocities, *strength effect* is more important than compressibility effect and this experimental result agrees qualitatively with numerical results of Sedgwick (1978):

$$\frac{P}{d_p} = 0.48 \cdot \left(\frac{\rho_p}{\rho_t}\right)^{0.54} \left(\frac{v_p}{(Y_t/\rho_t)^{1/2}}\right)^{0.47} \cdot \left(\frac{v_p}{c_t}\right)^{0.11}. \tag{9.18}$$

9.5 High Velocity Extension Fracture of Metallic Jets and Plates

Experiments show that velocity distribution in the metallic jet of a shaped charge possesses peculiar linearity and that *necking* and *fracture* occur after the jet stretches to a certain degree. Figure 9.8 shows spacial and temporary development of necking and fracture in a jet produced by certain shaped charges.

This experimental curve has front and rear portions having different variation trends, with the dividing point between those portions being roughly the jet element having velocity $v_j \approx 5$ km/s. Fractured jet elements lose penetration ability, making it necessary to explore mechanisms of *necking*, fragmentation and associated physical factors.

According to Cheng (1977) [10], the controlling mechanism of instability in the front portion of the jet is due to aerodynamics forces and instability in the rear portion of the jet is due to dynamic necking. For the rear portion, a simple similarity law is valid and this law is based on the following assumptions:

1. The jet has a slender body with initial diameter d_0 and linear velocity distribution.
2. Each element of the jet moves with constant velocity, so velocity gradient or strain rate is constant everywhere and denoted as $\left(\dfrac{d\varepsilon}{dt}\right)_0$.
3. Pressure in the jet is not high and jet density ρ can be regarded as constant.
4. Material can be regarded as being inelastic because tensile strain $\gg$ elastic strain. Yield strength of material Y can be considered to represent resistance to deformation and shrinkage of cross-section ψ can be considered to represent fracture property.

Affected by inertia and strength, the jet elongates gradually and breakage eventually occurs. Diameter of the broken element of the jet d_b is a function of parameters d_0, $\left(\dfrac{d\varepsilon}{dt}\right)_0$, ρ, Y and ψ:

$$d_b = f\left(d_0, \ \left(\frac{d\varepsilon}{dt}\right)_0, \ \rho, \ Y, \ \psi\right). \tag{9.19}$$

The first three parameters in the right hand side of (9.19) are mutually dependent and controlled by the principle of conservation of mass. Because each element of the jet moves with its own constant velocity, velocity of the jet element can be used as a coordinate to identify that element. For an element with length dl, velocity difference dv_j and mass dm, conservation of mass requires:

$$\frac{dm}{dv_j} = \text{constant}, \tag{9.20}$$

so:

$$\rho \cdot \frac{\pi d_0^2}{4} \cdot \frac{dl}{dv_j} = \text{constant.} \tag{9.21}$$

Because velocity gradient equals strain rate:

$$\frac{dv_j}{dl} = \left(\frac{d\varepsilon}{dt}\right)_0,$$

(9.21) can be:

$$\frac{\rho d_0^2}{\left(\dfrac{d\varepsilon}{dt}\right)_0} = \text{constant.} \tag{9.22}$$

It is reasonable to take ρ and $\dfrac{d_0^2}{\left(\dfrac{d\varepsilon}{dt}\right)_0}$ as independent parameters, and the latter is denoted as Ω:

$$\Omega = \frac{d_0^2}{\left(\dfrac{d\varepsilon}{dt}\right)_0}. \tag{9.23}$$

Thus, diameter of the broken element of jet d_b:

$$d_b = f(\Omega,\ \rho,\ Y,\ \psi). \tag{9.24}$$

Taking Ω, ρ and Y as a unit system, (9.24) reduces:

$$\frac{d_b}{\left[\left(\dfrac{Y}{\rho}\right)^{1/2} \cdot \Omega\right]^{1/3}} = f(\psi). \tag{9.25}$$

In (9.25), function f depends only on rate of shrinkage ψ. Therefore, f is constant for given material and denoted as c_d.

Breakage time can be expressed similarly:

$$\frac{t_b}{\left[\dfrac{\rho}{Y} \cdot \Omega\right]^{1/3}} = g(\psi) = c_t, \tag{9.26}$$

where function $g = $ constant depending on ψ for given material and denoted as c_t.

For the element of a jet with initial spatial position $z = b$ and velocity v_j, breakage occurs at time t_b and position z:

$$\begin{cases} t_b = c_t \cdot \left(\dfrac{\rho}{Y} \cdot \Omega\right)^{1/3}, \\[2mm] z = c_t \cdot \left(\dfrac{\rho}{Y} \cdot \Omega\right)^{1/3} \cdot v_j + b. \end{cases} \tag{9.27}$$

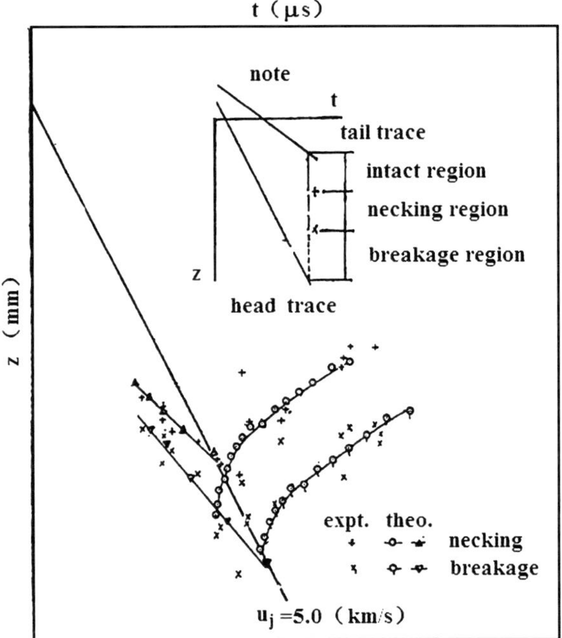

Fig. 9.8 Development of necking and breakage of shaped charge jet

Similar analytical approach can apply to time and position for necking, so results for necking and breakage of the rear portion of the jet ($v_j \leq 5$ km/s) may appear as solid curves in Fig. 9.8. This figure shows good agreement of experimental data, including necking points (+) and breakage points (×) .

Dimensional analysis is applicaple to *extension fracture of thin plates* with high extension velocity [11]. A typical example is the formation of fragments in high explosive shells. Impulsive fracture of thin plates results from inertia acting against strength, but tension in the plate is two-dimensional, unlike the one-dimensional tension in jet extension. Therefore, the relation of parameter Ω to velocity gradient velocity must be redefined.

It is assumed that at $t = 0$, there is an extending plate with thickness d_0, equal rates of strain $\left(\dfrac{d\varepsilon}{dt}\right)_0$ are in mutually orthogonal directions and each element of the plate moves with constant velocity in both directions perpendicular to the plate surface. Along both orthogonal directions, corresponding velocities of plate elements provide coordinates. For a plate element having the same length dl in both orthogonal directions, similar corresponding velocity difference dv_j and mass dm, conservation of mass requires:

$$\frac{dm}{dv_j dv_j} = \text{constant, or} \quad \frac{\rho d_0 dl dl}{dv_j dv_j} = \text{constant,} \tag{9.28}$$

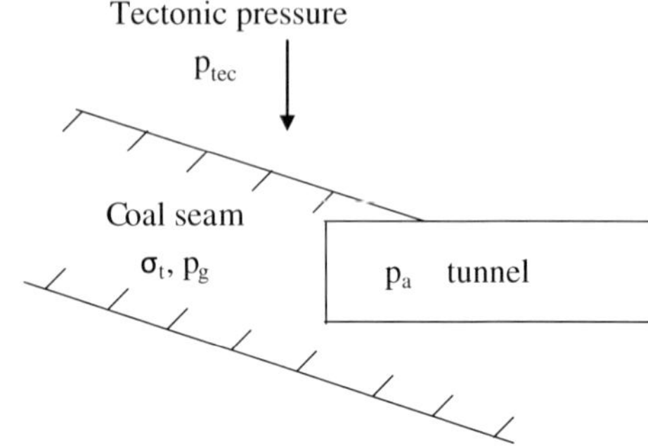

Fig. 9.9 Controlling parameters of coal and gas outburst

so:

$$\frac{\rho d_0}{\left(\dfrac{\mathrm{d}\varepsilon}{\mathrm{d}t}\right)_0^2} = \text{constant}$$

A new parameter Ω appears:

$$\Omega = \frac{d_0}{\left(\dfrac{\mathrm{d}\varepsilon}{\mathrm{d}t}\right)_0^2}. \tag{9.29}$$

Fracture time t_b and its concurrent thickness d_b:

$$\begin{cases} t_b = g(\psi) \cdot \left[\left(\dfrac{\rho}{Y}\right)^{1/2} \cdot \Omega\right]^{1/3}, \\[3mm] d_b = f(\psi) \cdot \left[\dfrac{Y}{\rho} \cdot \Omega\right]^{1/3}. \end{cases} \tag{9.30}$$

Result (9.30) differs from Grady's result [12]:

$$t_b = f\left(\frac{\rho v_0^2}{Y}\right) \cdot \frac{d_0}{(Y/\rho)^{1/2}}. \tag{9.31}$$

9.6 Coal and Gas Outburst Phenomenon Related to Coupled Two-Phase Medium

Coal and gas outburst threatens life and property in underground mining. During excavation, particularly soon after drilling and blasting, a coal face can suddenly collapse and fractured coal mass driven by pressurized gas can gush into tunnels (Fig. 9.9). In large-scale outburst, fractured coal masses can entirely fill a tunnel hundreds of meters long and fractured coal can bury miners and machinery. In such event, miners near the coal face cannot escape.

Coal and gas outburst is a difficult subject of analysis for reasons including complexity of geology. However, systematic laboratory study and theoretical analysis done by Cheng's research group [13, 14] in the 1980s and 1990s reveal the basic mechanism of such outbursts.

Data collected from large coal gas outbursts in China were examined in order to explore the mechanism of those outbursts through the use of order-of-magnitude analysis and dimensional analysis [12]. Outburst energy was shown to originate from potential energy of pressurized gas in coal seams. Critical conditions and dimensionless governing parameters associated with such outbursts were outlined.

9.6.1 Energy Origin of Outbursts

Table 9.1 shows field data related to nine serious outburst events. The first six rows show field data and the last five rows show data estimated on the basis of field data. Event no. 7 in Table 9.1 shows the greatest outburst in China, an event that happened in Sichuan Province in 1975. Intensity as defined by coal mass ejected was 12,780 tons and gas volume vented was roughly 1.4 million cubic meters.

Static pressure p at outburst position is estimated according to depth of coal seam H and mean density of covering stratum $\bar{\rho}$ (assumed to be about 2.5 g/cm^3). It is assumed that tectonic shear stress is close to 0 because of creep over long geologic period and assumed that tectonic stress is approximate to static pressure $p = \bar{\rho}gH$. If ρ is taken as being the density of coal seam (1.6 g/cm^3), then the volume of the ejected coal V_c can be estimated from outburst intensity M, so $V_c = \frac{M}{\rho}$. Mean radius of the cavity formed in seam R due to coal outburst can be estimated as $R = \left(\frac{M}{\pi \rho t}\right)^{1/2}$, where t = coal seam thickness.

Cheng estimated potential energy is stored in the ejected coal with volume V_c, but for the original state under tectonic pressure p and pore gas pressure p_g. Potential energy includes elastic energy of covering rock W_1, elastic energy of coal itself W_2 and internal energy of pore gas in coal in pressurized state W_g. To estimate elastic potential energy of cover W_1, it is assumed that the space occupied originally by the ejected coal is cylindrical and the estimate shows that $\frac{W_2}{W_1} \approx 2.0 \sim 0.2$. Estimate of the internal energy of pore gas W_g is based on assumptions that coal seam porosity = 0.1, the process is adiabatic and the adiabatic index of gas = 1.2.

The last row of Table 9.1 shows that elastic potential energy $W_1 \ll$ internal energy of pore gas W_g. Even for the most serious outburst no. 7, intensity is 12,800 tons and order of magnitude of elastic potential energy is smaller than that of pore gas. Therefore, it can be assumed that stored potential energy of pore gas in coal seams prone to outbursts exceeds by several orders of magnitude the elastic energy in both rock and coal. Clearly, pore gas pressure is the major driving force for coal and gas outburst.

Table 9.1 Nine serious outbursts in China

Event number	1	2	3	4	5	6	7	8	9
Depth H (m)	230	321	295	520	330	216	520	550	–
Thickness of coal seam t (m)	2.5	3.7	4.1	4.0	6.0	4.6	2.3	4.6	1.5
Inclination of coal seam (degree)	24	30	31	60	12	–	–	–	40
Pressure of pore gas p_g (kgf/cm^2)	24.6	–	18.5	36	16	17.5	8.0	>8.1	5.0
Outburst intensity M (tons)	1350	1000	2000	5270	1700	4500	12780	2800	1890
Volume of gas vented V (10^4m^3)	10.2	1.28	130	–	18	128	140	–	16.4
Tectonic pressure p (kgf/cm^2)	56.4	78.6	72.3	127	80.9	52.9	127	135	–
Gushed coal volume V_c (m^3)	844	625	1250	3290	1060	2810	7990	1750	1180
Cavity radius R (m)	10.4	7.34	9.85	16.2	7.51	4.41	35 ~ 29	11.0	15.8
R/t	4.1	2.0	2.4	4.1	1.3	0.96	18 ~ 10	2.4	11
Elastic energy W_1/Gas energy W_g	.006 ~ .004	–	.013 ~ .02	–	.018 ~ .004	.007 ~ .0001	.09 ~ .25	–	–

9.6.2 Critical Condition for Outburst

Laboratory simulations show two types of critical phenomena related to coal and gas outbursts. *Critical fracture* occurs if the combination of tectonic pressure and pore gas pressure is so strong in relation to coal seam strength that macro-fracture occurs. *Critical outburst* occurs if tectonic pressure or pore gas pressure increases further and synthetic intensity of tectonic pressure and gas pressure exceeds given strength of the coal seam so that there is continuous violent outburst of coal debris and gas.

Analysis of a real coal seam is complicated by fissures and faults. This discussion minimizes geological complications and the covering stratum and coal seam are regarded as being homogeneous. Because coal seam density influences rate of outburst only, density does not determine occurrence of fracture or outburst. In addition, because absorption and percolation of gas are much slower than fracture propagation, absorption and percolation effects can be ignored. Important governing parameters are:

Set 1. Loading: tectonic pressure p_t, pocket gas pressure p_g, atmospheric pressure p_a;

Set 2. Properties of materials: elastic constants of coal E_c and v_c, elastic constants of covering stratum E and v, tensile strength of coal σ_t, shear strength of coal σ_s, friction coefficient k at the roof or floor of coal seam, porosity of coal seam ε, adiabatic index of gas γ;

Set 3. Geometry: radius of excavation R, thickness of coal seam t.

Theoretical criterion for fracture or outburst:

$$f(p_t, p_g, p_a; E, v, E_c, v_c, \sigma_t, \sigma_s, k, \varepsilon, \gamma; R, t) = 0. \tag{9.32}$$

Taking p_a and t as a unit system reduces dimensional expression (9.32) to dimensionless relationship:

$$f\left(\frac{p_t}{p_a}, \frac{p_g}{p_a}; \frac{E}{p_a}, v, \frac{E_c}{p_a}, v_c, \frac{\sigma_t}{p_a}, \frac{\sigma_s}{p_a}, k, \varepsilon, \gamma; \frac{R}{t}\right) = 0. \tag{9.33}$$

Damage conditions depend mainly on whether or not synthetic intensity of overpressure $p_t - p_a$ or $p_g - p_a$ exceeds tensile strength of coal σ_t, so dimensionless damage condition can be:

$$f\left(\frac{p_t - p_a - \sigma_t}{p_a}, \frac{p_g - p_a - \sigma_t}{p_a}; \frac{E}{p_a}, v, \frac{E_c}{p_a}, v_c, \frac{\sigma_s}{\sigma_t}, k, \varepsilon, \gamma; \frac{R}{t}\right) = 0. \tag{9.34}$$

If materials are the same in model and prototype and both geometries are similar, critical conditions are:

$$f\left(\frac{p_t - p_a - \sigma_t}{p_a}, \frac{p_g - p_a - \sigma_t}{p_a}\right) = \text{constant}, \tag{9.35}$$

where constant in the right hand side of (9.35) has different values for macro-fracture or continuous outburst and the particular form of function f reflects a certain combination of the effects of tectonic pressure and pressure of pore gas.

9.6.3 Simulation Experiments in Coal Shock Tube

Since the 1980s, simulation experiments have been carried out using a shock tube containing two sections with different pressures. The section with higher pressure is filled with consolidated coal powders of fixed composition, density, porosity, strength and is charged with gas to pressure p_0. The section with lower pressure used to simulate a tunnel is exposed to environmental atmospheric pressure p_a. A diaphragm is positioned between these two sections so that sharp pressure drop $p_0 - p_a$ occurs between the two sections and simulates conditions in the coal seam and tunnel before fracture or outburst. Two critical conditions are of particular interest after the diaphragm breaks:

Critical condition 1. Determining how much overpressure $p_0 - p_a$ is needed to cause macro-fracture in the surface layer of a coal specimen helps provide a reference for preventing outburst.

Critical condition 2. Determining how much overpressure $p_0 - p_a$ is needed to push broken fragments of coal out of the shock tube helps provide a reference for dealing with the aftermath of outburst events.

Experiments show that increase in pore pressure p_0, relates to critical values p_{cr1} and p_{cr2}. When pore pressure p_0 reaches lower critical value p_{cr1}, macro-fracture appears in the surface layer of coal seam. With increase in p_0, the fracture region enlarges and with even more increase of p_0, broken coal debris separates from the coal seam and ejects from the shock tube. Until pore pressure p_0 reaches higher critical value p_{cr2}, the fracture region extends deeper into the coal seam. Broken coal debris is transported from the high pressure section to low pressure section that qualitatively corresponds to outburst in the coal tunnel.

There are studies of fracture occurrence caused by low pore gas pressures in which pressure p_0 is slightly greater than p_{cr1} [13]. Experiments show that, with increase in initial pore gas pressure, p_0 reaches p_{cr1} and a few discernible fissures form in the surface layer of the coal specimen. With further increase in p_0, thickness of the region with fissures increases and when p_0 increases to a certain value, a long crack runs transversely through the cross-section in the front part of the region having fissures. With more increase of p_0, a few fissures appear in front of the transverse crack. A second transverse crack follows with further increase in pressure. Other cracks occur periodically and follow the same process. Total depth at the front of the region with fissures L_D increases linearly with increase in p_0 (Fig. 9.10).

A breakage wave model [14] explains the development of the region with fissures. When the diaphragm breaks, an elastic unloading wave propagates into the high pressure section, pressure at the section end σ instantaneously drops

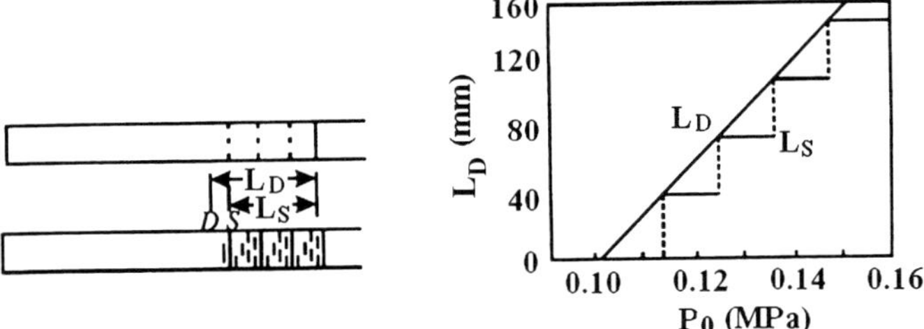

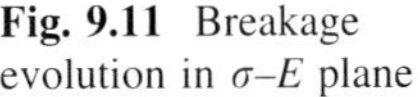

Fig. 9.10 Development of fissure region front L_D versus p_0

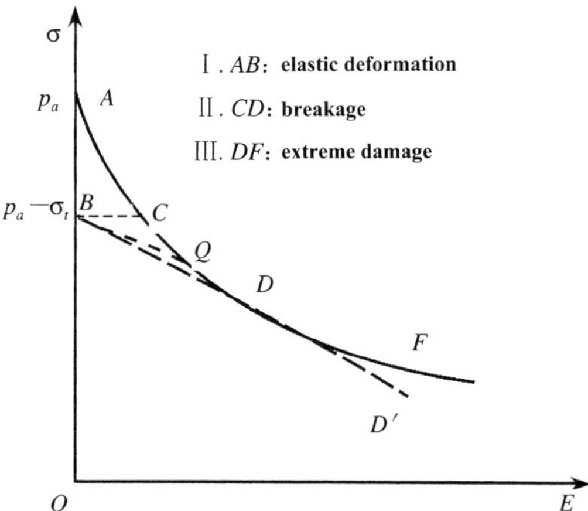

Fig. 9.11 Breakage evolution in $\sigma\text{-}E$ plane

from p_0 to atmospheric pressure p_a. If overpressure $p_0 - p_a >$ tensile strength of coal specimen σ_t, breakage occurs and propagates deep into the coal specimen. Measurements indicate that propagation velocity (~ 10 m/s) at the front of the breakage wave is much less than elastic wave velocity. Tensile tests suggest that, roughly speaking, the coal specimen is a rigid-brittle type material. Thus, the propagation velocity of an elastic wave in a coal specimen can be regarded as being infinite. For a particle of coal containing pore gas that is affected by an elastic tension wave, pressure σ instantaneously drops from p_0 to $p_0 - \sigma_t$. If $p_0 - \sigma_t$ is slightly greater than p_a, a local small region with fissures forms. If $p_0 - \sigma_t$ increases slightly, the region with fissures deepens. Such extension of the region with fissures can be regarded as being the propagation of a breakage wave. Hence, at the front of the breakage wave, pressure σ drops from the value ahead of the wave front $p_0 - \sigma_t$ to the value behind the wave front p and pressure drop $= p_0 - \sigma_t - p$. Correspondingly, porosity increases from initial value ε_0 to ε and it is reasonable to assume that pressure at the back side p follows isothermic variation. In Fig. 9.11, the curve of pressure σ versus logarithmic strain $E \left(= \ln \frac{1-\varepsilon_0}{1-\varepsilon}\right)$ reflects the rigid and brittle behavior of the coal specimen. This curve $ABCDF$ consists of a vertical part AB along ordinate σ axis from p_0 to $p_0 - \sigma_t$, representing the rigid property before fracturing and another curve part CDF, representing isothermic variation (Fig. 9.11). In the beginning of this experiment, pressure at the front of breakage wave is $p_0 - \sigma_t$ and pressure behind the front is p_a, so that the pressure drop across the wave front $p_0 - \sigma_t - p_a$ is the maximum for the whole process. Corresponding with the principle of conservation of energy, it seems reasonable to assume that state point (σ, E) behind the wave front is on the isothermic curve $ACQDF$ in Fig. 9.11. Therefore, across the wave front, the state point should jump from point B (ahead of the wave front) to point Q (behind the wave front) on the isothermic curve and the slope of the secant BQ is proportional to the square of the propagation velocity of the breakage wave. Possible maximum velocity of breakage wave is relevant to the slope of tangentBD because that tangent is the maximum of all slopes of possible secants. Following propagation of the breakage wave, pressure ahead of the wave front is constant $p_0 - \sigma_t$ and pressure behind the wave front rises from initial value p_a to σ. Correspondingly, the state point behind the wave front rises along the isothermic curve and pressure drop across the wave front decreases gradually. Pressure drop finally reaches to 0, the state point reaches point C and the breakage wave stops propagating. At that moment, a region with fissures of length L_D remains in the coal specimen.

Governing parameters used for determining the length of the region with fissures include:

Part 1. Coal specimen: density ρ, tensile strength σ_t;
Part 2. Pore gas: initial pressure p_0, initial porosity ε_0, dilatation index γ;
Part 3. Tube: diameter d, wall friction coefficient k;
Part 4. Air in the section with low pressure= pressure p_a.
Length of the region with fissures L_D is a function of above parameters:

$$L_D = f(\rho,\ \sigma_t;\ p_0,\ \varepsilon_0,\ \gamma;\ d,\ k;\ p_a). \qquad (9.36)$$

Taking ρ, d, and p_a as a unit system reduces dimensional expression (9.36) to dimensionless relationship:

$$\frac{L_D}{d} = f\left(\frac{\sigma_t}{p_a},\ \frac{p_0}{p_a},\ \varepsilon_0,\ \gamma,\ k\right). \qquad (9.37)$$

The first two independent variables should be combined to form maximum dimensionless pressure drop, $\frac{p_0 - p_a - \sigma_t}{p_a}$ at the initial moment, so:

$$\frac{L_D}{d} = f\left(\frac{p_0 - p_a - \sigma_t}{p_a},\ \varepsilon_0,\ \gamma,\ k\right). \qquad (9.38)$$

Considering the coal specimen to be a linear elastic and brittle material, relationship (9.38) is linear:

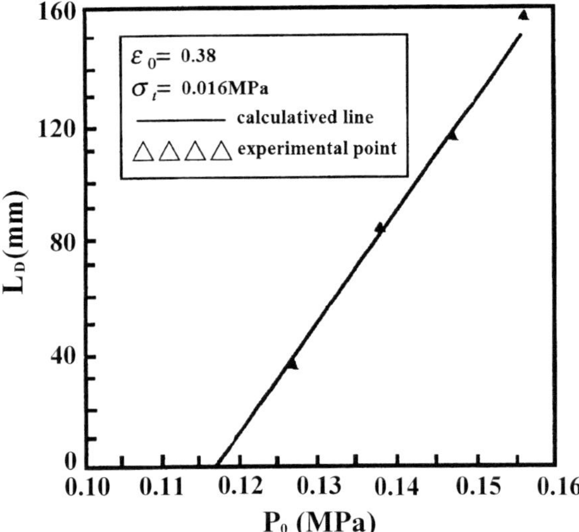

Fig. 9.12 Thickness of region with fissures x_m $(=L_D)$ versus initial pore gas pressure p_0

$$\frac{L_D}{d} \propto \frac{p_0 - p_a - \sigma_t}{p_a},\tag{9.39}$$

where proportional coefficient depends on ε_0, γ and k.

Experiments show linear relationship for given initial porosity ε_0, dilatation index of pore gas γ, diameter d and tube friction coefficient k. Figure 9.12 shows experimental and numerical results in the $L_D \sim p_0$ plane.

References

1. Gao, J.X., Le, M.K., Lu, D.Y., Xie, B.M.: A medeling law of projectile penetration. J. Ordnance **1**, 33–39 (1985). (in Chinese)
2. Sun, G.C., et al.: Modeling experiments of projectile penetration. Res. Rept., Institute of Mechanics, CAS. (1976). (in Chinese)
3. Taylor, G.I.: A Formulation of Mr. Tuck's conception of Munroe Jets (1943). Scientific Papers of G.I. Taylor, Cambridge University Press, New York, 358–362 (1963)
4. Cheng, C.M.: Preliminary analysis of high velocity jet penetration and related fundamental knowledge (1973) (in Chinese). Collected Works of Cheng C.M., pp. 224–294, Beijing: Science Press (2004)
5. Gao, J.X., Cheng, C.M., et al.: A medeling law of shaped charge jet penetration. Mechanics **1**, 1–10 (1974). (in Chinese)
6. Cheng, C.M., Tan, Q.M.: A dynamical analysis and a simplified model for high velocity jet penetration. Res. Rept., Institute of Mechanics, CAS. (1977). (in Chinese). Collected Works of Cheng C.M., pp. 295–351, Beijing: Science Press, (2004)
7. Zhu, Z.X.: Private Communication. (1977)
8. Wang, H., Bai, Y.L., Xia, M.F., Ke, F.J.: Spallation analysis with a closed trans-scale formulation of damage evolution. Acta. Mech. Sin. **20**(4), 400–407 (2004)
9. Kinslow, R.: High-velocity impact phenomena. Academic Press, New York City (1970)

10. Cheng, C.M.: Stability of shaped charge jet. Res. Rept., Institute of Mechanics, CAS. (1977). Collected Works of Cheng C M, pp. 303–320, Beijing: Science Press, (2004). (in Chinese)
11. Tan, Q.M.: Modeling laws in high velocity impacts. In: Wang, L.L. et al. (ed.) Advances in Impact Dynamics, Hefei: Publ. Co. of Univ. of Sci. and Tech. of China, pp. 303–320 (1992). (in Chinese)
12. Grady, D.E.: Fragmentation of rapidly expanding jets and sheets. Int. J. Impact Engng **5**, 285–292 (1987)
13. Cheng, C.M.: The mechanism of coal and gas outburst investigated by dimensional analysis and orders of magnitude. Collected Works of Cheng C.M., pp. 382–392, Beijing: Science Press, (2004). (in Chinese)
14. Tan, Q.M., Yu, S.B., Zhu, H.Q., Cheng, C.M.: Fracture of coal containing pressurized gas by sudden relieving. J. Ch. Coal Soc. **22**(5), 514–518 (1997). (in Chinese)

Chapter 10
Normalization in Mathematical Simulations

All the simulations described in earlier chapters deal with a single type of physical phenomena, so those simulations are called "physical simulations". In this chapter, we will discuss *mathematical simulation* that is different from the physical simulations. In mathematical simulation [1, 2], different types of physical phenomena can be described by common mathematical formulations and one type can be used to simulate another type. Since judgment of mathematical simulation based on intuition is usually suspect, it is necessary to adopt *normalization* to ensure correctness of mathematical simulation. Key in mathematical simulation is selecting proper units that form essential dimensionless quantities in order to simplify and approximate problems so that they can be solved reasonably. Therefore, dimensional analysis is also a powerful tool in mathematical simulation.

This chapter discusses normalization of functions, algebraic equations, ordinary differential equations and partial differential equations. Examples demonstrate principles and applications of normalization, showing some treatments that seem to be right but are actually wrong and other treatments that are reasonable and effective.

10.1 Normalization of Functions

Normalization of functions requires *selection of units* of independent variables and unit of function to make all the absolute values of the resultant dimensionless function and dimensionless derivatives of various orders less than 1.

It is supposed that velocity v is a function of distance x defined in domain $[0, x_0]$ (Fig. 10.1):

$$v = f(x), \quad 0 \leq x \leq x_0. \tag{10.1}$$

where function $f(x)$ involves physical parameters that reflect related physical essentials. The task is to find proper intrinsic scales or proper units of x and v from

Q.-M. Tan, *Dimensional Analysis*, DOI: 10.1007/978-3-642-19234-0_10,

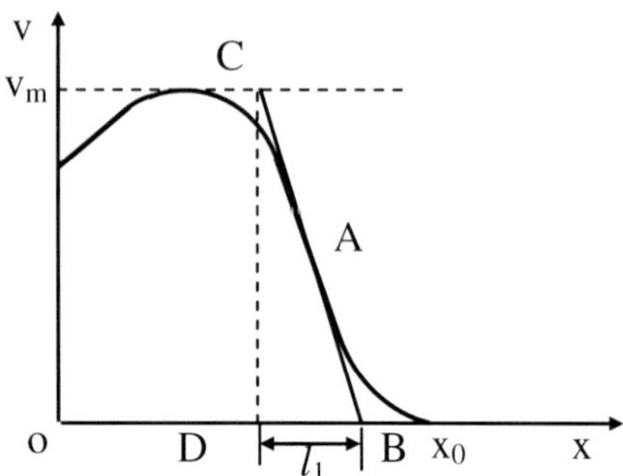

Fig. 10.1 Velocity $v \sim$ distance x

the function in (10.1). Selecting the unit of v is the easiest part of analysis because $|v|_m$, the maximum of the absolute value of $f(x)$, can be taken as a unit of v and dimensionless velocity v_* can be:

$$v_* = \frac{v}{|v|_m}. \tag{10.2}$$

Thus, $|v|_* \leq 1$ is in defined domain $0 \leq x \leq x_0$.

The unit of x is harder to find because that proper unit relates to magnitudes of derivatives of various orders. Regarding $\dfrac{dv}{dx}$, if absolute value of the slope of curve $v = f(x)$ has maximum $\left|\dfrac{dv}{dx}\right|_m$ at point A, a unit of x that is denoted as l can be taken from $|v|_m$ and $\left|\dfrac{dv}{dx}\right|_m$. The tangent of the curve at point A intersects abscissa x at point B and intersects horizontal line $v = |v|_m$ at point C. The projection of line segment CB on abscissa x is DB with length l_1, so:

$$l_1 = \frac{|v|_m}{|dv/dx|_m}. \tag{10.3}$$

The unit of x can equal l_1:

$$l = l_1. \tag{10.4}$$

The dimensionless independent variable x_* and dimensionless derivative $\left(\dfrac{dv}{dx}\right)_*$ can be:

$$x_* = \frac{x}{l}, \quad \text{and} \quad \left(\frac{dv}{dx}\right)_* = \frac{dv/dx}{|v|_m/l}. \tag{10.5}$$

Because:

$$\left|\left(\frac{dv}{dx}\right)_*\right| \leq \frac{|dv/dx|_m}{|v|_m/l} = 1, \tag{10.6}$$

the requirement that $\left|\left(\dfrac{dv}{dx}\right)_*\right| \leq 1$ in the defined domain $0 \leq x \leq x_0$ is satisfied.

Regarding $\dfrac{d^2v}{dx^2}$, if maximum absolute value of the second derivative of $v = f(x)$ is $\left|\dfrac{d^2v}{dx^2}\right|_m$, the ratio of $\left|\dfrac{dv}{dx}\right|_m$ and $\left|\dfrac{d^2v}{dx^2}\right|_m$ may be taken as a candidate unit of x and denoted as l_2:

$$l_2 = \frac{|dv/dx|_m}{|d^2v/dx^2|_m}. \tag{10.7}$$

For modification, it is assumed that the unit of x takes $\min(l_1, l_2)$:

$$l = \min(l_1, l_2). \tag{10.8}$$

The dimensionless independent variable x_* and dimensionless second derivative $\left(\dfrac{d^2v}{dx^2}\right)_*$ may be defined:

$$x_* = \frac{x}{l}, \quad \text{and} \quad \left(\frac{d^2v}{dx^2}\right)_* = \frac{d^2v/dx^2}{|v|_m/l^2}. \tag{10.9}$$

Because:

$$l^2 \leq l_1 \cdot l_2 = \frac{|v|_m}{|dv/dx|_m} \cdot \frac{|dv/dx|_m}{|d^2v/dx^2|_m} = \frac{|v|_m}{|d^2v/dx^2|_m},$$

then,

$$\left|\left(\frac{d^2v}{dx^2}\right)_*\right| = \frac{|d^2v/dx^2|}{|v|_m/l^2} \leq \frac{|d^2v/dx^2|_m}{|v|_m} \cdot \frac{|v|_m}{|d^2v/dx^2|_m} = 1, \tag{10.10a}$$

so $\left|\left(\dfrac{d^2v}{dx^2}\right)_*\right| \leq 1$ be surely in defined domain $0 \leq x \leq x_0$.

In addition:

$$\left(\frac{dv}{dx}\right)_* = \frac{dv/dx}{|v|_m/l} \quad \text{and} \quad \left|\left(\frac{dv}{dx}\right)_*\right| \leq 1. \tag{10.10b}$$

Adopting treatment similar to that used to define l_1 and l_2:

$$l_3 = \frac{|d^2v/dx^2|_m}{|d^3v/dx^3|_m}. \tag{10.11}$$

If selecting unit of x, $l = \min(l_1, l_2, l_3)$, dimensionless independent variable x_* can be

$$x_* = \frac{x}{l}, \tag{10.12a}$$

and dimensionless third derivative $\left(\dfrac{d^3 v}{dx^3}\right)_*$ can be

$$\left(\frac{d^3 v}{dx^3}\right)_* = \frac{d^3 v / dx^3}{|v|_m / l^3}. \tag{10.12b}$$

Therefore:

$$\left|\left(\frac{d^3 v}{dx^3}\right)_*\right| \leq 1. \tag{10.13}$$

and so on, allowing higher order derivatives to be handled.

Thus, proper intrinsic scales $|v|_m$ and l are selected as a unit system, ensuring that all absolute values of the dimensionless function v_* and various orders dimensionless derivatives $\left(\dfrac{d^n v}{dx^n}\right)_*$, $(n = 1, 2, 3\ldots)$ are < 1 and the requirement for normalization is satisfied.

10.2 Normalization of Algebraic Equations

A simple example concerns finding roots of a simultaneous system of algebraic equations:

$$\begin{cases} x + 10y = 21, \\ 5x + y = 7. \end{cases} \tag{10.14}$$

In the first equation, the coefficient of x is small compared to the coefficient of y, so it is tempting to ignore x and approximate the value of y:

$$y \approx 2.1,$$

Substituting the y approximation into the second equation approximates the value of x:

$$x \approx \frac{7 - 2.1}{5} = 0.98.$$

To check the approximation's validity, $(x, y) \approx (0.98, \ 2.1)$ is substituted into the first equation of (10.14), producing the ratio of the first term to the second term:

$$\frac{x}{10y} = \frac{0.98}{21} \approx 0.05 \ll 1.$$

This ratio validates $y \approx 2.1$. In fact, approximate roots $x \approx 0.98, y \approx 2.1$ are close to exact roots:

$$x = 1, \quad y = 2.$$

Another example concerns finding roots of the simultaneous system of algebraic equations:

$$\begin{cases} 0.01x + y = 0.1, \\ x + 101y = 11. \end{cases} \tag{10.15}$$

As in the previous example (10.14), considering the first equation in (10.15), the coefficient of x is much less than the coefficient of y and it is tempting to ignore the first term and approximate the value of y:

$$y \approx 0.1.$$

Substituting approximate $y \approx 0.1$ into the second equation produces approximate value of x:

$$x \approx 11 - 10.1 = 0.9.$$

To validate approximate solution $x \approx 0.9$, $y \approx 0.1$, the ratio of the first term to the second term in the first equation is:

$$\frac{0.01x}{y} = \frac{0.009}{0.1} = 0.09 \ll 1.$$

The approximation seems valid, but exact roots are:

$$x = -90, \quad y = 1.$$

Therefore, the "approximation" for y is inaccurate by a factor of 10 and the "approximation" for x is false because the wrong sign appears. Obviously, the approximation roots are erroneous.

To understand errors in the implementation of normalization requires analysis of the principle of normalization. Multiplying proper factors in both equations of (10.15) makes the right hand side equal 1 so as to normalize the equation system:

$$\begin{cases} \dfrac{1}{10}x + 10y - 1, \\ \dfrac{1}{11}x + \dfrac{101}{11}y = 1. \end{cases} \tag{10.16}$$

It is important to examine whether or not the first term in the first equation is much less than the right hand side $= 1$, and to notice that the relationship between two equations are simultaneous equations. In comparison with these two equations, coefficients of corresponding terms are close and there are relatively small differences, so (10.16) becomes:

$$\begin{cases} \dfrac{1}{10}x + 10y = 1, \\ \dfrac{1}{10}(1 + \varepsilon)x + 10(1 + \eta)y = 1, \end{cases} \tag{10.17}$$

where $\varepsilon = -\frac{1}{11}$ and $\eta = -\frac{9}{110}$ are small quantities in comparison with 1. Exact roots are:

$$\begin{cases} x = 10 \cdot \left(1 - \dfrac{1}{1 - \eta/\varepsilon}\right), \\[2ex] y = \dfrac{1}{10 \cdot (1 - \eta/\varepsilon)}. \end{cases}$$

In the expressions for x and y, common factor $1 - \eta/\varepsilon$ is in the denominator. If $\frac{\eta}{\varepsilon}$ is close to 1, the first term in the first equation $\frac{1}{10}x$ is much greater than the right hand side $= 1$, so $\frac{1}{10}x$ cannot be ignored. On the other hand, the ratio of the first term and the second term is:

$$\frac{\frac{1}{10}x}{10y} = \frac{1 - \frac{1}{1-\eta/\varepsilon}}{1 - \eta/\varepsilon} = 1 - \frac{\eta}{\varepsilon} - 1 = -\frac{\eta}{\varepsilon} = -\frac{9}{10}.$$

Therefore, the first term cannot be ignored.
Several conclusions appear:

1. It is unreasonable to compare only the coefficients of the x term and y term because the orders of magnitude of x and y are unknown beforehand. To determine which term can be ignored, it is important to begin by implementing normalization and then to compare the first term and the second term with the normalized right hand side $= 1$.
2. It is unreasonable to consider only a single equation in a simultaneous system of algebraic equations. Both equations should be considered together. Substituting exact roots into the equations produces:

$$\begin{cases} -9 + 10 = 1, \\[2ex] -\dfrac{90}{11} + \dfrac{101}{11} = 1. \end{cases}$$

The first term and second term belong to the same orders of magnitude and all are larger than the right hand sides $= 1$.

10.3 Normalization of Ordinary Differential Equations

Discussion of the well-known problem of an ejected projectile's trajectory illustrates how different approximations can produce different results, some of which are apparently right but actually wrong and some of which are reasonable and effective. Here, *gravity effect* is considered but air resistance is ignored.

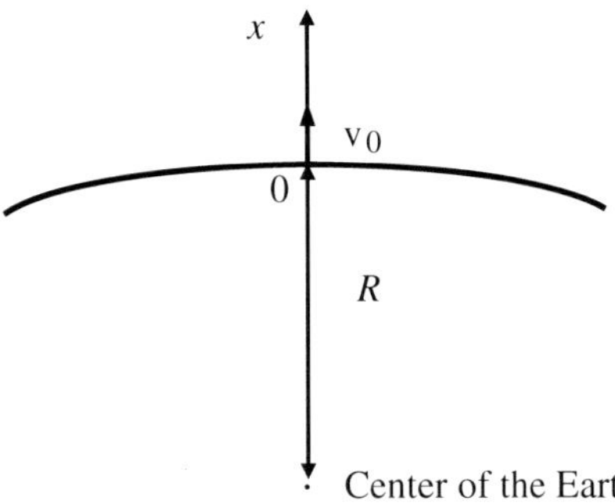

Fig. 10.2 Trajectory of projectile thrown-up

Gravitational force exerted on a projectile is conversely proportional to the square of the distance from the center of the Earth (Fig. 10.2). If the Earth's surface is taken as an origin of coordinate $x = 0$, x is positive in the upward direction and gravitational acceleration at the Earth's surface $= -g$, then the trajectory of a projectile $x = x(t)$ with initial velocity v_0 follows ordinary equation and initial conditions:

$$
\begin{cases}
\dfrac{d^2 x}{dt^2} = -g \cdot \dfrac{R^2}{(R+x)^2} \\[2mm]
t = 0: \quad x = 0, \quad \dfrac{dx}{dt} = v_0
\end{cases}
\tag{10.18}
$$

where R = radius of the Earth, g, R and v_0 are parameters in formulation (10.18). Projectile trajectory is:

$$
x = f(t;\ g,\ R,\ v_0).
\tag{10.19}
$$

There are at least two possible schemes for selecting a unit system. In Scheme 1, selecting R and v_0 as a unit system and $\frac{R}{v_0}$ as a characteristic time produces dimensionless expression for projectile trajectory:

$$
\frac{x}{R} = f\left(\frac{t}{R/v_0},\ \frac{v_0^2}{gR}\right).
$$

In Scheme 2, selecting R and g as a unit system and $\left(\frac{R}{g}\right)^{1/2}$ as a characteristic time produces dimensionless expression for projectile trajectory:

$$
\frac{x}{R} = f\left(\frac{t}{(R/g)^{1/2}},\ \frac{v_0}{(gR)^{1/2}}\right).
$$

Regardless of the unit system that is selected, the only dimensionless parameter is $\frac{v_0^2}{gR}$ or $\frac{v_0}{(gR)^{1/2}}$. In general, initial ejection velocity is so small that the projectile's distance from the Earth's surface $\ll$ radius of the Earth. Thus:

$$v_0 \ll (gR)^{1/2}.$$

The small parameter ε in the problem is taken as:

$$\varepsilon = \frac{v_0^2}{gR} \quad \text{or} \quad \varepsilon = \frac{v_0}{(gR)^{1/2}}. \tag{10.20}$$

Approximate solution can be found by a perturbation method, but the key is to *select a proper unit system*. In the following, adopting a unit selection based on two schemes searches for an approximate solution.

1. Scheme 1

Taking R as the unit of length, $\frac{R}{v_0}$ as the unit of time allows definition of dimensionless coordinate x_* and time t_*:

$$x_* = \frac{x}{R}, \quad \text{and} \quad t_* = \frac{t}{R/v_0}. \tag{10.21}$$

Dimensionless equation and initial conditions can be obtained:

$$\begin{cases} \varepsilon \dfrac{\mathrm{d}^2 x_*}{\mathrm{d}t_*^2} = -\dfrac{1}{(x_* + 1)^2}, \\[2mm] t_* = 0: \quad x_* = 0, \quad \dfrac{\mathrm{d}x_*}{\mathrm{d}t_*} = 1 \end{cases} \tag{10.22}$$

where ε is small parameter $\frac{v_0^2}{gR}$.

It is assumed that $x_*(t_*)$ can be expanded in a series of powers of small parameter ε:

$$x_*(t_*) = x_{*0}(t_*) + \varepsilon x_{*1}(t_*) + \varepsilon^2 x_{*2}(t_*) + \cdots. \tag{10.23}$$

Substituting $x_*(t_*)$ into the equation and the initial condition produces:

$$\begin{cases} \varepsilon \dfrac{\mathrm{d}^2 x_{*0}}{\mathrm{d}t_*^2} + \varepsilon^2 \dfrac{\mathrm{d}^2 x_{*1}}{\mathrm{d}t_*^2} + \cdots = -\dfrac{1}{(x_{*0} + 1)^2}\left(1 - 2\varepsilon \dfrac{x_{*1}}{x_{*0} + 1} + \cdots\right) \\[2mm] t_* = 0: \quad x_{*0} + \varepsilon x_{*1} + \cdots = 0, \quad \dfrac{\mathrm{d}x_{*0}}{\mathrm{d}t_*} + \varepsilon \dfrac{\mathrm{d}x_{*1}}{\mathrm{d}t_*} + \cdots = 1 \end{cases}$$

Comparing the zeroth order terms of ε in the left hand side and right hand side of the equation and initial condition:

$$\begin{cases} -\dfrac{1}{(x_{*0} + 1)^2} = 0, \\[2mm] t_* = 0: \quad x_{*0} = 0, \quad \dfrac{\mathrm{d}x_{*0}}{\mathrm{d}t_*} = 1. \end{cases} \tag{10.24}$$

Obviously, no solution exists.

2. Scheme 2

Taking R as the unit of length and $(R/g)^{1/2}$ as the unit of time, dimensionless coordinate x_* and time t_* can be defined:

$$x_* = \frac{x}{R}, \quad \text{and} \quad t_* = \frac{t}{(R/g)^{1/2}}. \tag{10.25}$$

It is assumed that $x_*(t_*)$ can be expanded in a series of powers of small parameter $\varepsilon = \frac{v_0}{(gR)^{1/2}}$:

$$x_*(t_*) = x_{*0}(t_*) + \varepsilon x_{*1}(t_*) + \varepsilon^2 x_{*2}(t_*) + \cdots. \tag{10.26}$$

Substituting $x_*(t_*)$ into the equation and the initial condition, $x_{*0}(t_*)$ satisfies:

$$\begin{cases} \dfrac{\mathrm{d}^2 x_{*0}}{\mathrm{d}t_*^2} = -\dfrac{1}{(x_{*0}+1)^2}, \\[2mm] t_* = 0: \quad x_{*0} = 0, \quad \dfrac{\mathrm{d}x_{*0}}{\mathrm{d}t_*} = 0. \end{cases} \tag{10.27}$$

Obviously, the solution is not positive:

$$x_{*0} < 0.$$

Therefore, the solution is wrong.

Schemes 1 and 2 fail to solve by means of the perturbation method because selected units of length and time do not represent characteristics of the problem. In particular, length R is not representative and neither of the orders of magnitude of x_* or t_* is 1.

However, it is possible to select a proper characteristic length as a unit of length. Because maximum height $x_m \ll$ the radius of the Earth R, gravitational acceleration can be regarded as an approximate constant g in the whole range of the trajectory. Thus, maximum height is roughly:

$$x_m \approx \frac{v_0^2}{2g}.$$

Range of time variation t can also be estimated and the time when the projectile arrives at the top of the trajectory is roughly:

$$t_m \approx \frac{v_0}{g}.$$

Taking $\frac{v_0^2}{g}$ as the unit of length and $\frac{v_0}{g}$ as the unit of time produces dimensionless coordinate and dimensionless time:

$$x_* = \frac{x}{\left(v_0^2/g\right)}, \quad \text{and} \quad t_* = \frac{t}{(v_0/g)}. \tag{10.28}$$

Hence, there are dimensionless equation and initial conditions:

$$\begin{cases} \dfrac{d^2 x_*}{dt_*^2} = -\dfrac{1}{(1 + \varepsilon x_*)^2}, \\[3mm] t_* = 0: \quad x_* = 0, \quad \dfrac{dx_*}{dt_*} = 1 \end{cases} \tag{10.29}$$

where, ε is a small parameter $\dfrac{v_0^2}{gR}$.

It is assumed that $x_*(t_*)$ can be expanded in a series of powers of small parameter ε:

$$x_*(t_*) = x_{*0}(t_*) + \varepsilon x_{*1}(t_*) + \varepsilon^2 x_{*2}(t_*) + \cdots . \tag{10.30}$$

Thus, $x_{*0}(t_*)$ satisfies the equation and initial condition:

$$\begin{cases} \dfrac{d^2 x_{*0}}{dt_*^2} = -1, \\[3mm] t_* = 0: \quad x_{*0} = 0, \quad \dfrac{dx_{*0}}{dt_*} = 1. \end{cases} \tag{10.31}$$

The approximate solution is:

$$x_{*0} = t_* - \frac{t_*^2}{2}. \tag{10.32a}$$

Solution (10.32a) reduces to dimensional form:

$$x_0 = v_0 t - \frac{gt^2}{2}, \tag{10.32b}$$

Form (10.32b) is a well-known trajectory for constant gravitational acceleration. This expression provides the zeroth approximation of the solution of the projectile problem with variable gravitational acceleration (10.18).

Full understanding of a problem requires reviewing interpretation of each dimensionless parameter from the viewpoint of physics. In the present case, the unique parameter is $\varepsilon = \dfrac{v_0^2}{gR}$, the well-known Froude number. In previous chapters, the Froude number represented the ratio of effects of *inertia* and *gravity*. Here, ε can be regarded as the ratio of two lengths $\dfrac{v_0^2/g}{R}$. The denominator is the radius of the Earth R and the numerator $\dfrac{v_0^2}{g}$ is an estimate of the maximum height achieved by a projectile with ordinary initial velocity v_0.

10.4 Normalization of Partial Differential Equations

This section discusses a case involving heat conduction in solids and a case involving *boundary layer* flow in hydrodynamics.

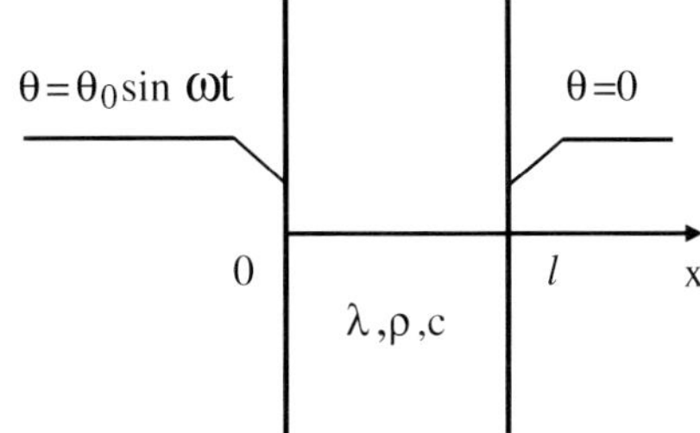

Fig. 10.3 One-dimensional heat conduction

10.4.1 One-Dimensional Heat Conduction

Figure 10.3 shows a one-dimensional heat conduction problem. Solid body thickness $= l$ and coordinate x is taken along the direction of thickness. Both surfaces of the solid body are assigned given temperature T or temperature difference θ $(= T - T_0)$. At one surface $x = l$, given temperature is constant:

$$T = T_0,$$

so temperature difference:

$$\theta = 0.$$

At another surface $x = 0$, given temperature difference is periodic:

$$\theta = \theta_0 \sin \omega t,$$

where $\theta_0 =$ amplitude and $\omega =$ angular frequency. Predicting temperature difference in the solid body after a relatively long time depends on boundary conditions but not on initial conditions.

Temperature difference θ satisfies partial differential equation and boundary conditions:

$$\begin{cases} \dfrac{\partial \theta}{\partial t} = \dfrac{\lambda}{\rho c} \cdot \dfrac{\partial^2 \theta}{\partial x^2}, \\ x = 0: \quad \theta = \theta_0 \sin \omega t, \\ x = l: \quad \theta = 0, \end{cases} \tag{10.33}$$

where $\lambda =$ heat conduction coefficient, $\rho =$ density, $c =$ specific heat, $\theta_0 =$ amplitude of temperature difference and $\omega =$ angular frequency. Parameters in the equation and boundary conditions are $\frac{\lambda}{\rho c}$, θ_0, ω and l, therefore variation in temperature difference can be:

$$\theta = f\left(x, \ t; \ \frac{\lambda}{\rho c}, \ \theta_0, \ \omega, \ l\right). \tag{10.34}$$

Taking θ_0, ω and l as a unit system produces;

$$\frac{\theta}{\theta_0} = f\left(\frac{x}{l},\ \omega t;\ \frac{\lambda}{\rho c \omega l^2}\right). \tag{10.35}$$

Selecting l as characteristic length, and $\frac{1}{\omega}$ as characteristic time allows definition of dimensionless independent variables and dependent variable:

$$x_* = \frac{x}{l},\quad t_* = \omega t,\quad \theta_* = \frac{\theta}{\theta_0}. \tag{10.36}$$

Dimensionless equation and boundary conditions are:

$$\left\{ \begin{array}{l} \dfrac{\partial \theta_*}{\partial t_*} = \dfrac{\lambda}{\rho c \omega l^2} \cdot \dfrac{\partial^2 \theta_*}{\partial x_*^2}, \\[2mm] x_* = 0: \quad \theta_* = \sin t_*, \\[2mm] x_* = 1: \quad \theta_* = 0. \end{array} \right. \tag{10.37}$$

Mathematical formulation (10.37) has only one-dimensionless parameter, Fourier number $\frac{\lambda}{\rho c \omega l^2}$.

It is interesting to discuss the case of $\frac{\lambda}{\rho c \omega l^2} \ll 1$. If material of the solid body is given, or $\frac{\lambda}{\rho c}$ is given, there are two possible explanations for $\frac{\lambda}{\rho c \omega l^2} \ll 1$. One explanation is for given angular frequency ω of the surface temperature difference when thickness of the solid body l is variable, $\frac{\lambda}{\rho c \omega l^2} \ll 1$ is equivalent to $l \gg \left(\frac{\lambda}{\rho c \omega}\right)^{1/2}$. Another explanation is for the given thickness of the solid body l when angular frequency ω is variable, $\frac{\lambda}{\rho c \omega l^2} \ll 1$ is equivalent to $\omega \gg \frac{\lambda}{\rho c l^2}$. Thus, temperature variation in the solid body occurs in a thin layer near the surface $x = 0$. If the thickness of that thin layer is denoted as δ, relative thickness $\frac{\delta}{l}$ is a small quantity that the thin layer may be called a *temperature boundary layer*. Therefore, thickness of the solid body l is unimportant, thickness of the temperature boundary layer δ is essential and δ should be taken as characteristic length instead of l. Taking δ as a unit of length allows redefinition of the special dimensionless coordinate:

$$x_* = \frac{x}{\delta}, \tag{10.38a}$$

while still taking:

$$t_* = \omega t,\quad \text{and}\ \theta_* = \frac{\theta}{\theta_0}. \tag{10.38b}$$

There are now dimensionless equation and boundary conditions:

$$\left\{ \begin{array}{l} \dfrac{\partial \theta_*}{\partial t_*} = \dfrac{\lambda}{\rho c \omega \delta^2} \cdot \dfrac{\partial^2 \theta_*}{\partial x_*^2}, \\[2mm] x_* = 0: \quad \theta_* = \sin t_*, \\[2mm] x_* \rightarrow \infty: \quad \theta_* = 0. \end{array} \right. \tag{10.39}$$

Considering the properties of the thermal source, $\theta_* = \sin t_*$ at $x_* = 0$ produces $|\theta_*| \leq 1$ and $\left| \dfrac{\mathrm{d}\theta_*}{\mathrm{d}t_*} \right| \leq 1$. These thermal source properties can be used to estimate magnitudes on both sides of the differential equation. Obviously, the order of magnitude of the left hand side of equation $\frac{\partial \theta_*}{\partial t_*}$ is 1. If δ is properly selected, the order of magnitude of $\frac{\partial^2 \theta_*}{\partial x_*^2}$ in the right hand side is also 1. To do this, $\frac{\lambda}{\rho c \omega \delta^2}$ can equal 1 and δ can be:

$$\delta = \left(\frac{\lambda}{\rho c \omega} \right)^{1/2}, \tag{10.40}$$

In fact, such δ is an intrinsic scale of length in the problem. For the mathematical problem:

$$\begin{cases} \dfrac{\partial \theta_*}{\partial t_*} = \dfrac{\partial^2 \theta_*}{\partial x_*^2}, \\[2mm] x_* = 0 : \quad \theta_* = \sin t_*, \\[2mm] x_* \to \infty : \quad \theta_* = 0. \end{cases} \tag{10.41}$$

The usual solution for θ_* is:

$$\theta_* = \frac{\theta}{\theta_0} = \exp\left(-\sqrt{\frac{\rho c \omega}{2\lambda}} \cdot x \right) \cdot \sin\left(\omega t - \sqrt{\frac{\rho c \omega}{2\lambda}} \cdot x \right). \tag{10.42}$$

This solution satisfies the boundary condition at $x = l$ because:

$$\exp\left(-\sqrt{\frac{\rho c \omega}{2\lambda}} \cdot l \right) \ll 1.$$

From this case of heat conduction, it is clear that if $\frac{\lambda}{\rho c \omega l^2}$ is sufficiently small, a thinner plate can be adopted to simulate temperature variation in a thicker plate.

10.4.2 Boundary Layer of Viscous Flows

At the start of the twentieth century, the Wright brothers made the first powered controlled flight and their planes were manufactured on the basis of their experiences. At the end of the nineteenth century, theoretical hydrodynamics was advanced enough to solve linear Euler equation for inviscid fluid flows but not to solve nonlinear Navier–Stokes equations for viscous fluid flows. The theory of inviscid fluid dynamics demonstrates that neither resistance nor lift acts on a body moving uniformly in an inviscid fluid. Of course, this theory is contrary to fact, so it is referred to as *d'Alembert's paradox*.

Based on experimental observations, *Prandtl* (1904) [3] pointed out that viscosity of fluids has significant effect only in a thin layer in the neighborhood of the

Fig. 10.4 Viscous flow past
two-dimensional airfoil

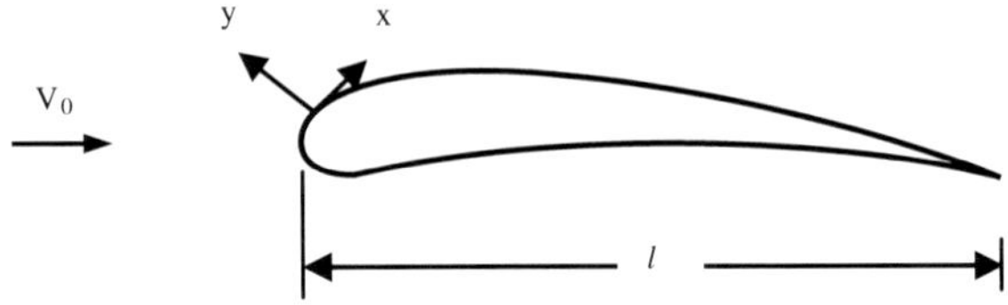

body (*boundary layer*), so viscous friction can be ignored in the region outside this layer. *Prandtl* ingeniously applied order of magnitude analysis and dimensional analysis to examine parts played in a set of conservation equations by all terms, including inertia term, viscosity term and pressure term. These equations were simplified and approximate solutions enabled easier research. Based on Prandtl's boundary layer theory, causes of resistance and lift of an airplane were clearly revealed. On one hand, viscous stress produced in the boundary layer induces resistance; on the other hand, viscous shear motion in the boundary layer induces circulatory flow around the airplane that produces lift. Prandtl's boundary layer theory was a great leap in the development of hydrodynamics. Since then, the theory of hydrodynamics has become a theoretical basis for aircraft design.

In this section, boundary layer equations for *laminar* flows are derived via the principle of normalization and Blasius's self-similar solution for *viscous flow past a flat plate* (1908) is introduced.

Figure 10.4 shows steady flow past a two-dimensional aerofoil of incompressible viscous fluid with velocity v_0. A natural coordinate is applied, with origin set at the stagnation point of the head of the airfoil. The x-axis extends along the surface of the aerofoil with the y-axis perpendicular to the surface. Conservation equations and boundary conditions can be expressed:

$$\left\{ \begin{aligned} &\text{Mass conservation: } \frac{\partial u}{\partial x} + \frac{\partial v}{\partial y} = 0 \\[6pt] &\text{Momentum conservation: } u\frac{\partial u}{\partial x} + v\frac{\partial u}{\partial y} = -\frac{1}{\rho}\frac{\partial p}{\partial x} + \frac{\mu}{\rho}\left(\frac{\partial^2 u}{\partial x^2} + \frac{\partial^2 u}{\partial y^2}\right), \\[6pt] &\qquad\qquad\qquad\quad u\frac{\partial v}{\partial x} + v\frac{\partial v}{\partial y} = -\frac{1}{\rho}\frac{\partial p}{\partial x} + \frac{\mu}{\rho}\left(\frac{\partial^2 v}{\partial x^2} + \frac{\partial^2 v}{\partial y^2}\right); \\[6pt] &\text{Boundary conditions:} \\ &\quad \text{At the aerofoil surface: } f_w(x, y, l, l', \ldots) = 0 : u = 0, v = 0, \\ &\quad \text{At infinity: } (x, y) \to \infty : u = v_0, v = 0. \end{aligned} \right. \tag{10.43}$$

In mathematical formulation (10.43), unknowns are velocity components u, v, and pressure p and parameters are density ρ, fluid viscosity coefficient μ, incoming fluid velocity v_0 and geometrical parameters l, $l', \ldots$ that describe airfoil shape $f_w(x, y, l, l', \ldots) = 0$. Clearly, velocity distribution depends on:

$$
\begin{cases}
u = f_u(x,\ y;\ \rho,\ \mu;\ v_0;\ l,\ l',\ldots), \\
v = f_v(x,\ y;\ \rho,\ \mu;\ v_0; l,\ l',\ldots).
\end{cases}
\tag{10.44}
$$

Taking ρ, v_0, and l as a unit system, (10.44) reduces to dimensionless expression:

$$
\begin{cases}
\dfrac{u}{v_0} = f_u\!\left(\dfrac{x}{l},\ \dfrac{y}{l},\ \dfrac{\rho v_0 l}{\mu},\ \dfrac{l'}{l},\ldots\right), \\[2mm]
\dfrac{v}{v_0} = f_v\!\left(\dfrac{x}{l},\ \dfrac{y}{l},\ \dfrac{\rho v_0 l}{\mu},\ \dfrac{l'}{l},\ldots\right).
\end{cases}
\tag{10.45}
$$

When Reynolds number Re $\left(= \frac{\rho v_0 l}{\mu}\right)$ is less than its critical value $(\text{Re})_{cr}$, fluid flow in the boundary layer is laminar. In the case of the boundary layer in a flat plate, $(\text{Re})_{cr} \approx 10^6$.

According to requirement for the principle of normalization, selecting correct unit of length is key. It is proper to select the chord of the aerofoil l as a unit of length for measuring x but it is improper to measure y because, as Prandtl shows, *viscosity effect* only plays a significant part in a very thin boundary layer. It is just as well to assume that boundary layer thickness is δ and $\delta \ll l$, so using δ to measure y is proper. Because of the peculiarity of this problem, different units of length are selected in directions x and y. Nevertheless, characteristic velocity v_0 can serve temporarily as a unit of velocity.

Dimensionless independent and dependent variables can be defined:

$$
\begin{cases}
x_* = \dfrac{x}{l}, & y_* = \dfrac{y}{\delta}, \\[2mm]
u_* = \dfrac{u}{v_0}, & v_* = \dfrac{v}{v_0}, & p_* = \dfrac{p}{\rho v_0^2}.
\end{cases}
\tag{10.46}
$$

Clearly, in the dimensionless variables in (10.46), orders of magnitude of x_* and $u_* = 1$ and order of magnitude of v_* is unknown. If δ is selected properly, the order of magnitude of $y_* = 1$. Examining and comparing terms presented in the dimensionless equations allows selection of proper δ, leading to analysis of the order of magnitude of v_*.

Substituting dimensionless variables in (10.46) into the system of equations (10.43) produces:

$$
\begin{cases}
\dfrac{\partial u_*}{\partial x_*} + \dfrac{l}{\delta}\dfrac{\partial v_*}{\partial y_*} = 0; \\[4mm]
u_*\dfrac{\partial u_*}{\partial x_*} + \dfrac{l}{\delta}v_*\dfrac{\partial u_*}{\partial y_*} = -\dfrac{\partial p_*}{\partial x_*} + \dfrac{\mu}{\rho v_0 l}\left(\dfrac{\partial^2 u_*}{\partial x_*^2} + \left(\dfrac{l}{\delta}\right)^2\dfrac{\partial^2 u_*}{\partial y_*^2}\right), \\[4mm]
u_*\dfrac{\partial v_*}{\partial x_*} + \dfrac{l}{\delta}v_*\dfrac{\partial v_*}{\partial y_*} = -\dfrac{l}{\delta}\dfrac{\partial p_*}{\partial y_*} + \dfrac{\mu}{\rho v_0 l}\left(\dfrac{\partial^2 v_*}{\partial x_*^2} + \left(\dfrac{l}{\delta}\right)^2\dfrac{\partial^2 v_*}{\partial y_*^2}\right).
\end{cases}
\tag{10.47}
$$

In estimating the order of magnitude of each term in the equation of conservation of mass, the order of magnitude of $\frac{\partial u_*}{\partial x_*} = 1$, so the order of magnitude of $\frac{l}{\delta}\frac{\partial v_*}{\partial y_*} = 1$. Thus, to make the order of magnitude of y_* equal 1, the order of magnitude of v_* can be estimated as $\frac{\delta}{l}$:

$$O(v_*) = \frac{\delta}{l} \ll 1. \tag{10.48}$$

To analyze the equation of conservation of momentum in the direction of x, because all orders of magnitude of u_*, x_* and y_* are 1 and the order of magnitude of v_* is $\frac{\delta}{l}$, the orders of magnitude of both terms in the left hand side are 1. Considering that the viscous term in the right hand side and the inertia term in the left hand side should have the same order of magnitude, their orders of magnitude should be 1. However, considering the bracket factor of the viscous term, the order of magnitude of $\frac{\partial^2 u_*}{\partial x_*^2}$ is 1 and the order of magnitude of $\left(\frac{l}{\delta}\right)^2\frac{\partial^2 u_*}{\partial y_*^2}$ is $\left(\frac{l}{\delta}\right)^2 \gg 1$. Because these orders of magnitude are so different, $\frac{\partial^2 u_*}{\partial x_*^2}$ can be ignored. Based on analysis of the order of magnitude of the viscous term, the order of magnitude of δ can be estimated:

$$O(\delta) = \frac{l}{(\rho v_0 l/\mu)^{1/2}} = \frac{l}{\mathrm{Re}^{1/2}}. \tag{10.49}$$

The estimate in (10.49) leads to estimate of the order of magnitude of $\frac{\partial p_*}{\partial x_*}$:

$$O\left(\frac{\partial p_*}{\partial x_*}\right) = 1. \tag{10.50}$$

To analyze the equation of conservation of momentum in the direction of y, it is clear that all orders of magnitude of two terms in the left hand side are $\frac{\delta}{l}$, while the order of magnitude of the viscous term in the right hand side is $\frac{1}{\mathrm{Re}} \cdot \frac{l}{\delta}$, which is equal to $\frac{\delta}{l}$. The order of magnitude of the term of pressure gradient in the right hand side can be estimated:

$$O\left(\frac{l}{\delta} \cdot \frac{\partial p_*}{\partial y_*}\right) = \frac{\delta}{l}.$$

Thus, the order of magnitude of $\frac{\partial p_*}{\partial y_*}$ is:

$$O\left(\frac{\partial p_*}{\partial y_*}\right) = \left(\frac{\delta}{l}\right)^2. \tag{10.51}$$

According to the principle of normalization, if $\frac{\partial p_*}{\partial y_*}$ is compared with $\frac{\partial p_*}{\partial x_*}$ $\left(O\left(\frac{\partial p_*}{\partial x_*}\right) = 1\right)$, it is reasonable to ignore $\frac{\partial p_*}{\partial y_*}$:

$$\frac{\partial p_*}{\partial y_*} = 0. \tag{10.52}$$

To sum up, simplified dimensionless equations are:

$$\begin{cases} \dfrac{\partial u_*}{\partial x_*} + \dfrac{l}{\delta}\dfrac{\partial v_*}{\partial y_*} = 0; \\[2ex] u_*\dfrac{\partial u_*}{\partial x_*} + \dfrac{l}{\delta}v_*\dfrac{\partial u_*}{\partial y_*} = -\dfrac{\partial p_*}{\partial x_*} + \dfrac{\mu}{\rho v_0 l}\left(\dfrac{l}{\delta}\right)^2\dfrac{\partial^2 u_*}{\partial y_*^2} \\[2ex] 0 = \dfrac{\partial p_*}{\partial y_*}. \end{cases} \tag{10.53}$$

The set of equations (10.53) shows that pressure in the boundary layer is practically constant along the direction perpendicular to the wall. That pressure may be assumed to be equal to that at the outer edge of the boundary layer, where value is determined by outer inviscid flow. Therefore, pressure in the boundary layer is merely a function of x, denoted by $p(x)$, that can be found by solving the outer flow using Euler's equations for inviscid fluid.

Dimensionless equations in (10.53) reduce to dimensional simplified Navier–Stokes equations, known as *Prandtl's boundary layer equations*:

$$\begin{cases} \dfrac{\partial u}{\partial x} + \dfrac{\partial v}{\partial y} = 0; \\[2ex] u\dfrac{\partial u}{\partial x} + v\dfrac{\partial u}{\partial y} = -\dfrac{1}{\rho}\dfrac{dp}{dx} + \dfrac{\mu}{\rho}\dfrac{\partial^2 u}{\partial y^2}. \end{cases} \tag{10.54}$$

A complete mathematical formulation of steady flow in the boundary layer appears:

1. Equations:

$$\begin{cases} \dfrac{\partial u}{\partial x} + \dfrac{\partial v}{\partial y} = 0; \\[2ex] u\dfrac{\partial u}{\partial x} + v\dfrac{\partial u}{\partial y} = -\dfrac{1}{\rho}\dfrac{dp}{dx} + \dfrac{\mu}{\rho}\dfrac{\partial^2 u}{\partial y^2}, \end{cases} \tag{10.55a}$$

2. Boundary conditions:

$$\begin{cases} \text{At aerofoil surface:} \\ f_w(x,\, y,\, l,\, l',\ldots) = 0: \quad u = 0,\ v = 0, \\ \text{At infinity:} \quad y \to \infty: \quad (u,\, v) \to (U,\, V), \end{cases} \tag{10.55b}$$

where $(U,\, V)$ and p are velocity and pressure of the outer inviscid flow field.

A complicated viscous flow problem can be solved in two steps:

Step 1. Potential flow $U(x, y)$ and $V(x, y)$ outside the boundary layer are solved by using Euler's equations. Pressure distribution $p(x)$ can then be obtained by means of Bernoulli's theorem.

Step 2. Boundary layer equations are solved.

A simple example of the application of the boundary layer equations is afforded by the viscous flow over a flat plate with length l that is discussed by *Blasius* (1908). Because *viscosity effect* is confined in a thin layer and $\frac{\delta}{l} \ll 1$, parameter l does not considerably affect flow for a large portion of the layer and affects only a small portion near the rear end of the plate. It is reasonable to assume that the solution for the boundary layer of a flat plate with semi-infinite length is an approximate solution for a flat plate with length l. In general, dimensionless velocity distribution for a flat plate with length l can be expressed:

$$\begin{cases} \dfrac{u}{v_0} = f_u\left(\dfrac{x}{l}, \dfrac{y}{\delta}, \dfrac{\rho v_0 l}{\mu}\right), \\[3mm] \dfrac{v}{v_0} = f_v\left(\dfrac{x}{l}, \dfrac{y}{\delta}, \dfrac{\rho v_0 l}{\mu}\right), \end{cases} \tag{10.56}$$

where δ can be replaced by $\dfrac{l}{\mathrm{Re}^{1/2}}$, so:

$$\begin{cases} \dfrac{u}{v_0} = f_u\left(\dfrac{x}{l}, \dfrac{y}{[(\mu l)/(\rho v_0)]^{1/2}}, \dfrac{\rho v_0 l}{\mu}\right), \\[3mm] \dfrac{v}{v_0} = f_v\left(\dfrac{x}{l}, \dfrac{y}{[(\mu l)/(\rho v_0)]^{1/2}}, \dfrac{\rho v_0 l}{\mu}\right). \end{cases} \tag{10.57}$$

For a flat plate with semi-infinite length, velocity distribution should not relate to l or to $\frac{\rho v_0 l}{\mu}$. Therefore, independent variables $\frac{x}{l}$ and $\frac{y}{[(\mu l)/(\rho v_0)]^{1/2}}$ cannot exist independently but can exist as a single independent variable by the ratio of $\frac{y}{[(\mu l)/(\rho v_0)]^{1/2}}$ to $\left(\frac{x}{l}\right)^2$, i.e.,:

$$\frac{y}{[(\mu l)/(\rho v_0)]^{1/2}} \bigg/ \left(\frac{x}{l}\right)^{1/2} = \left(\frac{\rho v_0}{\mu}\right)^{1/2} \cdot \frac{y}{x^{1/2}}. \tag{10.58}$$

In this way, boundary layer flow is a *self-similar flow* that depends on a single independent variable $\frac{y}{\sqrt{x}}$, i.e.,:

$$\begin{cases} \dfrac{u}{v_0} = f_u\left(\sqrt{\dfrac{\rho v_0}{\mu}} \cdot \dfrac{y}{\sqrt{x}}\right), \\[3mm] \dfrac{v}{v_0} = f_v\left(\sqrt{\dfrac{\rho v_0}{\mu}} \cdot \dfrac{y}{\sqrt{x}}\right). \end{cases} \tag{10.59}$$

Thus, solving a difficult problem related to partial differential equations is simplified by solving a much easier problem related to ordinary differential equations.

References

1. Lin, C.C., Segel, L.A.: Mathematics Applied to Deterministic Problems in Natural sciences. Macmillan Publ. Co., Inc., New York (1974)
2. Kline, S.J.: Similitude and Approximation Theory. McGraw-Hill, New York (1965)
3. Prandtl, L.: Fluid motion with very small friction (Über Flüssigkeitbewegung bei sehr kleiner Reibung), at the Heidelberg Mathematical Congress in 1904. In: Schlichting, H. (ed.) Boundary Layer Theory. McGraw-Hill, New York (1979)

Index

Q.-M. Tan, *Dimensional Analysis*, DOI: 10.1007/978-3-642-19234-0,
© Springer-Verlag Berlin Heidelberg 2011

Printed by Books on Demand, Germany